community forest management

A Casebook from India

Joe Human and Manoj Pattanaik

facing page: Tree nursery, Kesharpur village

First published by Oxfam GB in 2000

© Oxfam GB 2000

ISBN 0 85598 439 2

A catalogue record for this publication is available from the British Library.

All rights reserved. Reproduction, copy, transmission, or translation of any part of this publication may be made only under the following conditions:

- With the prior written permission of the publisher; or
- With a licence from the Copyright Licensing Agency Ltd., 90 Tottenham Court Road, London W1P 9HE, UK, or from another national licensing agency; or
- For quotation in a review of the work; or
- Under the terms set out below.

This publication is copyright, but may be reproduced by any method without fee for teaching purposes, but not for resale. Formal permission is required for all such uses, but normally will be granted immediately. For copying in any other circumstances, or for re-use in other publications, or for translation or adaptation, prior written permission must be obtained from the publisher, and a fee may be payable.

Available from the following agents:
USA: Stylus Publishing LLC, PO Box 605, Herndon, VA 20172-0605, USA
tel: +1 (0)703 661 1581; fax: +1(0)703 661 1547; email: styluspub@aol.com
Canada: Fernwood Books Ltd, PO Box 9409, Stn. 'A', Halifax, N.S. B3K 5S3, Canada
tel: +1 (0)902 422 3302; fax: +1 (0)902 422 3179; e-mail: fernwood@istar.ca
India: Maya Publishers Pvt Ltd, 113-B, Shapur Jat, New Delhi-110049, India
tel: +91 (0)11 649 4850; fax: +91 (0)11 649 1039; email: surit@del2.vsnl.net.in
K Krishnamurthy, 23 Thanikachalan Road, Madras 600017, India
tel: +91 (0)44 434 4519; fax: +91 (0)44 434 2009; email: ksm@md2.vsnl.net.in
South Africa, Zimbabwe, Botswana, Lesotho, Namibia, Swaziland: David Philip Publishers, PO Box 23408, Claremont 7735, South Africa
tel: +27 (0)21 64 4136; fax: +27(0)21 64 3358; email: dppsales@iafrica.com
Tanzania: Mkuki na Nyota Publishers, PO Box 4246, Dar es Salaam, Tanzania
tel/fax: +255 (0)51 180479, email: mkuki@ud.co.tz
Australia: Bush Books, PO Box 1958, Gosford, NSW 2250, Australia
tel: +61 (0)2 043 233 274; fax: +61 (0)2 092 122 468, email: bushbook@ozemail.com.au

Rest of the world: contact Oxfam Publishing, 274 Banbury Road, Oxford OX2 7DZ, UK.
tel. +44 (0)1865 311 311; fax +44 (0)1865 313 925; email publish@oxfam.org.uk

Designed and typeset by Oxfam GB
Printed by Information Press, Eynsham, Oxford.

Oxfam GB is a registered charity, no. 202 918, and is a member of Oxfam International.

Front cover photograph: Rajnikant Yadav/Oxfam

Contents

Acknowledgements v

Introduction 1

- 1 The years of struggle 3
- 2 The wider picture 22
- 3 India's forest peoples 38
- 4 The people fight back 58
- 5 BOJBP: from birth to maturity 73
- 6 Problems and challenges 98
- 7 The elements of success 127
- 8 Lessons learned 139

Appendix 1: *Thengapalli*: a resource for primary schools 145

Appendix 2: BOJBP publications 154

Appendix 3: Update on BOJBP and the *Mahasangha* 156

Notes 160

Further reading 169

Index 171

Acknowledgements

The origins of this book go back to 1991, when I first visited *Bruksha O' Jeevar Bandhu Parishad* ('BOJBP' – Friends of Trees and Living Beings) with Oxfam colleagues from Bhubaneswar. Astonished by the achievements of the movement, which were so clear for all to see in the form of hills re-clothed in forest, I wrote in their visitors' book:

> *November 18th has been an inspiring day. Thank you for sharing it with me. In turn I will share it with Oxfam friends and supporters in Britain. In that way I will try to reach many people with the message 'Trees are our life' – your message, which is the message for us all.*

Over the years since that day I have kept my promise and have told the story literally hundreds of times to a very wide variety of people, and am still telling it, to Oxfam volunteers and staff, primary-school children, sixth formers, school teachers, undergraduate and graduate students, environmental activists, women's groups, church organisations, and foresters. Thousands have heard it.

When it was proposed that I should write the story, I felt the need to work with someone closer to the forests of Orissa than I was. After much discussion, Manoj Pattanaik, from the Bhubaneswar-based Regional Centre for Development Co-operation, was asked to work with me on the project, and he agreed. This book is the outcome, therefore, of collaboration between us, with Manoj working in India and myself in Oxford, in touch all the time by e-mail. (Without it the project would have been much more complicated and a good deal slower.)

While technology has eased the process of writing, people have been essential to it. Indeed the book would have been impossible without the help of an enormous number of individuals. We want particularly to acknowledge our debt to Joginath Sahoo, Udayanath Katei (*Bapa*), and Professor Narayan Hazari, who have played such crucial roles in the formation and development of BOJBP, and who generously gave us of their time. But we interviewed numerous others, whose views helped to shape and give substance to the book and without whom it simply could not have been written in the way we wanted to write it. They are Anand Charan Acharya, Bhagaban Acharya, Ramesh Kumar Barad, Kanaka Barik, Biswanath Basantia, Kalandi Charan Behera, Gadadhara Bhatta, Mrs. Buli, Puna Dei, Bhobani Dora, S. Madhusudan Dora, Dhoba Gadu, Ms. Geeta, Manoj Hazary, Anil Kumar Jena, Sunderi Jena, Susant Jena, Uttama Jena, Purna Chandra Khandual, Digambar Mahapatra, Pitabasa Mahapatra, Kumudini Maharana, Rumi Maharana, Dambarudhara Majhi, Gokula Maharana, Sudarshan Malla, Ms. Mamata, Subas Chandra Mishra, Ajay Mohanty, Bankanidhi Mohanty, Govind Mohanty, Balabadhra Mohapatra, Binodini Mohapatra, Mamata Mohapatra, Purna Chandra Mohapatra, Subasini Mohapatra, Alaka Nanda, Kanchanabala Nayak, Hazari Parida, Baikuntha Pattanaik, Gopal Pattanaik, Kulamani Pattanaik, Pradip Pattanaik, Purna Chandra Pattanaik, Balaram Pradhan, Banchhanidhi Pradhan, Bhagaban Pradhan, Bula Pradhan, Jaladhara Pradhan, Kaushalya Pradhan, Malati Pradhan, Neema Pradhan, Kalpana Senapati, Bansidhar Sahoo, Bisnuprasad Sahoo, Kailash Sahoo, Lalita Sahoo, Nrusingha Sahoo, Purna Chandra Sahoo, Taramani Sahoo, Ms. Satyabhama, Bhagaban Subudhi, Anupama Swain, Aparna Swain, Pratima Swain, Prativa Swain, Santilata Swain, Shashi Swain, and Suresh Chandra Swain.

We are especially grateful to Priya Nilimani, who facilitated many of our interviews and who was helpful in numerous other ways, clarifying issues for us and giving many useful insights. Without him our work would have been very much more difficult. Bhikhyakari Hazary and Prahallad Sathua, the present Secretary of BOJBP, were also very helpful, and we are most grateful to them, as we are to Manoj Rathy and Niranjan, who gave us much support during our visits. At the *Mahasangha,* four people were generous with their time, and with their insights into the development of the *Mahasangha* itself and BOJBP. They are Barna Baibhav Panda ('*Bubu*'), Lakhmidara Balia, Bhagaban Dash, and Rabi Parida. We also acknowledge the help given to us by members of three sister organisations, *Sabuja Jeevan, Ratnamala Jungle Surakhya Committee,* and *Sulia Paribesa Parishad.* In addition to all these there were others, too numerous to mention, who made us very welcome and who hosted us on our visits.

We also want to thank Sriramappa, Oxfam's former Regional Representative, and Sarthak Pal, Project Officer in Bhubaneswar, Professor A. B. Mishra of Sambalpur University, and Joginath Sahoo for reading the draft and making many useful comments. I would like to add Sabita Banerji and Abhijit Bhattacharjee in Oxfam House in Oxford, and John Gwynn in Oxfam's Delhi office, who have been personally very supportive to me.

I would like to thank Dylan Theodore, who saw the power and universal relevance of the story of Friends of Trees and Living Beings, and had the vision to translate it into a brilliant resource – *Thengapalli* – for teachers and children in Britain. I would also like to thank Chris Kilby of the Cedars School, Veronica Lucas and children of Thornhill Junior School, and Martine Cox and children at King's Copse Primary School, all in Hampshire, who, among so many, have taken *Thengapalli* into their hearts and into ways of thinking about the environment, and who shared with me their responses to it.

The photographs in the book have come from Oxfam's photo-library and from the *Thengapalli* pack. We thank Hampshire County Council for permission to use those photos from the pack that are not in our library. Almost all are from the studio of Rajendra Shaw, of the Centre of Development Communications in Hyderabad. One, the cover photo, is by Rajnikant Yadav, another Oxfam-commissioned photographer. We are grateful to Kate Theodore for permission to use the map of Keshapur which she drew for the *Thengapalli* pack.

We thank Oxford University Press in Delhi for permission to use the map, Figure 5 (Chapter 3), which is taken from *Village Voices, Forest Choices,* edited by Mark Poffenberger and Betsy McGean. We thank Ashish Publishing House, Delhi, for permission to quote extensively from *Environmental Management in India,* Volume II, edited by P.K. Saparo.

I am grateful to Catherine Robinson, my editor at Oxfam Publishing, for her meticulous attention to detail and consistent support.

Finally I would like to thank, Jill, my wife, without whom this project would never have been completed.

Whatever help and support we have received from all those named and unnamed, the end product is our responsibility alone, and any errors of omission and commission, interpretation and inference lie entirely with us.

Joe Human
Oxford
July 2000

Introduction

This is the story of a remarkable organisation: Friends of Trees and Living Beings (*Bruksha O' Jeevar Bandhu Parishad* – BOJBP). The story has been told orally hundreds of times, and over the years has been heard by thousands of people. We have chosen to write it now for several reasons. We want it to reach more people, to inspire and motivate them. We want to tell it more fully, to consider not only the achievements of BOJBP, but also the considerable challenges now facing it, so that lessons may be learned. We also want to set the story in a wider context of environmental activism in India, within the framework of Indian forest policy and legislation, relating it to the situation of other forest-dependent peoples. The story, its context, and the lessons to be learned from it are told in eight chapters. There are three appendices.

Chapter 1, 'The years of struggle', sets the scene, describes the State of Orissa and Kesharpur, the village where BOJBP was born, looks at the state of the local forests, and considers what it was that stirred the village to action.

Chapter 2, 'The wider picture', describes the wealth of India's forests, and their importance in sustaining millions of people. It examines the framework of laws and policies and how they work against the interests of forest-dependent people. It considers the current state of India's forests, and describes attempts to halt their degradation and to devise effective ways of managing and enhancing forest resources.

Chapter 3, 'India's forest peoples', looks at the various groups of people who rely on forests, at the richness of their knowledge and skills, and at the impact of 'progress' on their lives and livelihoods, through deforestation, big 'development' projects, and the erosion of their rights. It considers the rate of displacement and examines the particular effects of environmental destruction on the lives of women.

Chapter 4, 'The people fight back', describes other movements in Orissa and India that have challenged the destruction of the environment and organised people in its and their own defence. Some have been successful, others have been crushed and died out. The current picture is one of powerful activism and widespread networking, primed to respond to any development that may threaten people's lives and livelihoods.

Chapter 5, 'BOJBP: from birth to maturity', continues the story of Friends of Trees and Living Beings from the early 1980s to the early 1990s, years when the organisation spread from its 'mother area' of 22 villages to more than 300, through the propagation of 'sister organisations'. These were dynamic years, when a range of techniques of persuasion was developed with great effectiveness, when members established a seed bank and nursery, evolved systems of governance, framed rules, attempted to inculcate a 'green culture', and tried to develop a gender-focused perspective. The chapter also looks at the role of Oxfam, at problems that arose from over-rapid growth, and at inter-communal conflicts.

Chapter 6, 'Problems and challenges', brings the story up to date, considers the formation of the *Mahasangha*, initially a campaigning and co-ordinating arm of Friends of Trees, and looks at more serious attempts to involve women in the movement. It considers the challenge of developing sustainable and strategic forest-management systems, and looks at BOJBP's response to Social Forestry and Joint Forest Management. The chapter concludes with an examination of the current problems facing the movement.

Chapter 7, 'The elements of success', looks at the quality of the organisation's leadership, communal cohesion and the spirit of voluntarism, the sharing of responsibilities, the question of rights, and the effectiveness of the strategies of persuasion and winning friends and influencing people, of involving children and spreading 'greenness'. It considers the importance of external links, including that with Oxfam. Finally it examines the significance of the organisation's stand against Joint Forest Management.

Chapter 8, 'Lessons learned', draws out the importance of the organisation's story and experience for policy makers, bilateral donors, Oxfam, other forest-protecting communities, and non-government organisations.

Appendix 1, *Thengapalli*, tells how Friends of Trees and Living Beings has inspired an enormously important resource for environmental education in Britain and brought the story alive for thousands of primary-school children, to learn from and be inspired by. Appendices are often not read, but this one is important for those interested in education, particularly if they are concerned to understand how the North can learn from the South.

Appendix 2 lists all the publications produced so far by BOJBP, and Appendix 3 brings the story of BOJBP and the *Mahasangha* up to date, with the latest available information as the book goes to press in July 2000.

There are two routes through this book. One is simply to read the story itself, as told in Chapters 1, 5, 6, 7, and 8. The other, which has the benefit of providing the wider context and important background, is to read it straight through. The reader must choose.

The years of struggle I

'Fifty years ago there was dense forest around our village. Binjhagiri was full of wild animals. Tigers used to reside in this forest ... By 1960–62 the forest was completely degraded. The trouble was that after Independence in 1947, people became careless and destroyed the forest.'
(Bankanidhi Mohanty, a member of the General Body of Friends of Trees, interviewed for this book)

The context: Orissa

Orissa, the Indian State where this story is located, lies on the eastern side of the country, facing the Bay of Bengal. Compared with some of its giant neighbours, it is of modest size. Its area is a little larger than that of England and Wales, comprising 4.7 per cent of India's land-mass. In 1998 its population was estimated to be a little over 35 million, just 3.7 per cent of India's total.

In a country of much poverty, Orissa is one of India's poorest States. Wherever the poverty-line is drawn – high or low – Orissa has the dubious distinction of having a greater proportion of people living below the line than any other major Indian State. Its per capita income is not only much lower than the national average, but the gap is widening. Nearly two-thirds of Orissa's workers are engaged in agriculture, either as farmers or as farm labourers. For most of them, living off the land is a precarious occupation, since more than three-quarters of their holdings are so small as to be considered marginal and uneconomic.

In June 1998, using figures from 1995, the Population Foundation of India published a compound Human Development Index which placed Orissa fourth from the bottom in a list of 16 major Indian States. The Foundation's tables show that in that year Orissa had the highest infant-mortality rate of any major Indian State (103 per thousand live births). It also came second from bottom in a Gender Health Index table, which measures the differences between males and females in terms of educational attainment, infant mortality, and life expectancy at birth.

Among the reasons often cited for Orissa's poor showing in these and other tables – as if this were somehow an excuse – is the fact that it has large proportions of its population from the Scheduled Castes and Scheduled

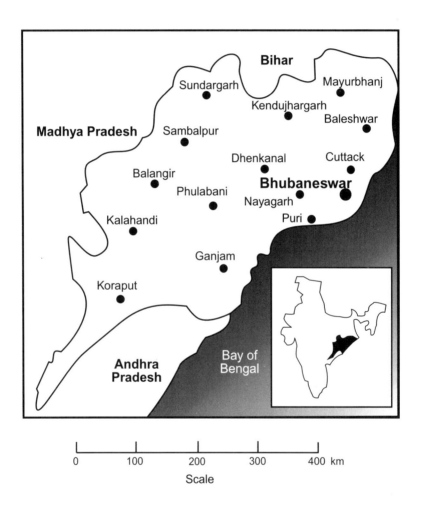

Figure 1: The State of Orissa, showing Nayagarh and principal towns.

Tribes[1] (above 16 per cent and more than 22 per cent respectively). Scheduled Castes are the former 'Untouchables', those whom Gandhi called *harijans* or 'children of God', but who call themselves *dalits*, meaning 'the oppressed'. Their numbers nationally are about 150 million (16 per cent of the population). Scheduled Tribes are the descendants of the original inhabitants of the country. They call themselves *adivasi*, meaning the 'original people'. Nationally they number more than 70 million (7.5 per cent of the population). Over the centuries both groups have suffered at the hands of the more powerful castes, and their suffering continues today. In Orissa, as elsewhere, they, more than any other groups, have suffered great exploitation, which has left most of them socially and economically more disadvantaged and much weaker than they were even at Independence in 1947. In the early 1980s, for example, it was reckoned that no fewer than 96 per cent of 3.5 million bonded labourers in Orissa were from Scheduled Caste and Scheduled Tribe communities, living in virtual slavery.[2]

While the Scheduled Castes and Scheduled Tribes are both vulnerable groups, in Orissa it is the situation of the Scheduled Tribe communities that is particularly precarious. One reason for this is that their homelands contain immense natural resources, needed by the modern Indian economy for wealth creation – that is wealth creation for India, not for tribal communities. Their ground has been dug up for its minerals, their rivers dammed for power and irrigation, their soils ploughed for commercial agriculture, and their forests felled for their timber.

Commenting on the worsening plight of Orissa's tribal peoples in the context of the liberalisation of the Indian economy, Oxfam Bhubaneswar's Strategic Plan, 1996–2000, describes them as increasingly 'helpless victims', subject to aggressive 'looting[3] of ... their habitats and environment'. One stark statistic above all others reveals the extent of that habitat destruction. In 1971-72, forests covered 43.5 per cent of Orissa. By 1993–94, coverage had declined to 36.5 per cent. In absolute terms, that is a loss in little more than 20 years of 10,900 sq km of forest, equivalent to just over half the area of Wales. By any measure this is an astounding rate of destruction. While most of this forest was occupied by tribal peoples, as we shall see in the section which follows, it was not exclusively so. Other communities have also lost their forests at the hands of outside interests, and at their own hands too.

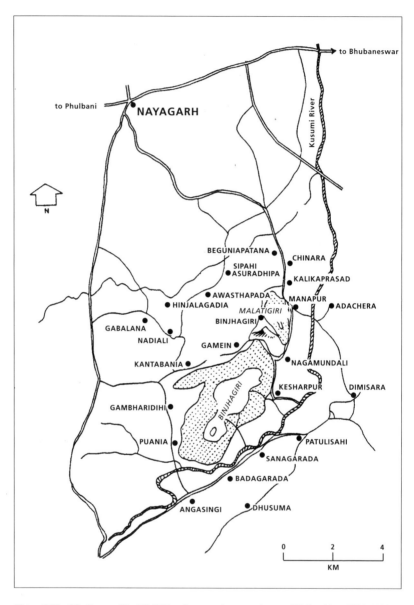

Figure 2: The 22 villages of the BOJBP 'mother area', grouped around Binjhagiri and Malatigiri. (Note: 23 settlements comprise 22 'revenue villages'.)

Kesharpur

Around the village

Kesharpur, the little village where the remarkable movement of *Bruksha O' Jeevar Bandhu Parishad* (Friends of Trees and Living Beings, BOJBP) was born, lies outside the tribal belt of Orissa, some 95 km west of Bhubaneswar, the State capital, and some 65 km inland from the coast. The nearest town is Nayagarh,[4] the District headquarters, to which it is connected by a 13-km single-lane, tarmac road laid on the red earth. Arriving in Kesharpur from Nayagarh, one would find nothing obvious about the village to explain why such a remarkable movement should have started here. To get a clue to it all, you would have to cast your eyes to the hill, Binjhagiri, which rises behind the village like a whale from the sea. Measuring just under 3 km long, just over 1 km wide, and 280 m high, Binjhagiri covers about 360 ha. Immediately to the north lies a small hill, Malatigiri, covering about 110 ha. Both are part of the broken ranges of the Eastern Ghats, which lie on this side of peninsular India. Thirty years ago both hills, like so many others in these ranges, were bare. Now they are clothed with rich forests. In the middle distance lie other whale-backed hills, rising from the plain, some much bigger than Binjhagiri. They too are clothed with forests, evidence of the remarkable work of BOJBP.

Botanists describe the forests here as 'dry mixed deciduous',[5] which means that most trees shed their leaves in the dry season. And although in India people often refer to forests like this as 'jungle' (*jangala*), these are not at all what people in the northern hemisphere would think of as jungles. In density they are more like temperate deciduous woodlands, although they are quite different in terms of the variety and type of trees and shrubs within them. Whereas a patch of temperate woodland would typically have at the most one dozen tree species, these have well over one hundred.

According to Indian forest law, the forests on Binjhagiri and Malatigiri are designated 'Undemarcated Protected Forests', locally called *Khesra*. (See Chapter 2 for a full explanation of the categorisation of India's forests.) This designation means that local villagers, while they do not own the forests (even though they may regenerate and protect them), have *bona fide* rights to their products. The map (Figure 2) shows that around Binjhagiri, within one kilometre of its foot, are nine villages, of which Kesharpur is one. Farther out lie another 13. In the past, to some degree, all 22 villages regarded the forests on these hills as 'theirs', in the sense that they saw their products – timber, fuelwood, fruits, fodder, medicinal herbs, and so on – as theirs to harvest and use. But the nine villages nearest to the hills have always been most dependent on them.

Community forest management

Figure 3: Sketch map from the *Thengapalli* resource pack (see Appendix 1). The tightly packed 'High Street' has a temple at each end and a meeting platform in the middle. The BOJBP compound is on the Nayagarh road out of the village to the north. The regenerated spring (see Figure 8) in the forest, just to the west of the compound, flows into the Kusumi, with some water diverted into the tanks. The small settlement near the largest tank is that of Kesharpur's ten Scheduled Caste families, still not fully integrated into the community.

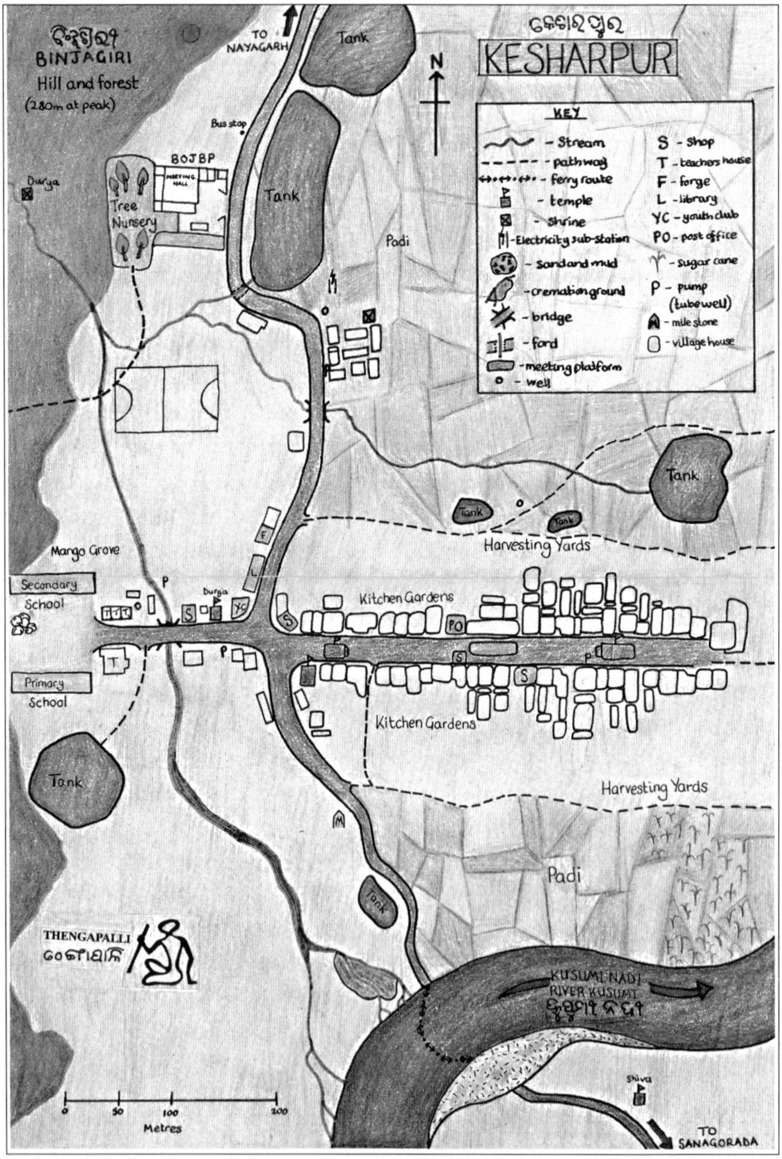

The years of struggle

Figure 4: 'High Street', Kesharpur. Apart from the stone house in the foreground, the houses are constructed of traditional local materials – thatch and earth, mixed with cow-dung. The concrete meeting platform in the centre is used during the day to dry crops.

Flowing gently past Kesharpur is the Kusumi river, fed by small streams that rise in springs on the flanks of the hill. To reach the villages south of Kesharpur, one crosses the river by a ford, but in the dry season the water recedes. Recently a bridge has been built to take the road on a by-pass around the village, but the connecting road itself has yet to be completed. *'And that won't happen for ten years!'*, according to Purna Mohapatra, one of BOJBP's committee members.

The village and the villagers

As the sketch map (Figure 3) and the photograph (Figure 4) show, the village is compact, with houses and a few shops tightly packed on either side of the main street. Living here are some 125 families – about 650 people. Most of their homes are made of earth, earthen bricks, timber, and thatch. A few are made of concrete and stone and others of laterite blocks.[6] Each house has a small verandah at the front and a vegetable patch at the rear, beyond which are the harvesting yards where crops are brought for threshing. The richer households have electricity, which was brought to the village in 1962. In the centre of the village is a platform, around which people gather for meetings and on which in the daytime they dry their crops in the sun. Around the platform, carts are often parked, and cows and water buffaloes stand quietly chewing the cud. At each end of the street lies a small temple. Halfway down there is a post office, and there are also two lock-up, kiosk-type shops, selling basic commodities like soap and matches. The street is always busy – not with cars, but with people doing their daily chores, coming and going from the fields, the wells, and the temples.

9

Just beyond the main part of the village are the two schools, primary and middle, whose children have played such a crucial role in the BOJBP movement. All round the village, hemmed in by the hill to the west and the river to the south and east, are fields growing mainly paddy (rice) and black gram (a pulse) for subsistence, and sugar cane for cash. The soils of these fields are mainly ancient laterites – the red earth that covers so much of peninsular India. But here and there are patches of black cotton soil, which turns to a glutinous mud in the monsoons. These rains, which fall mainly from June to October, average 1500 mm but vary enormously from year to year, and there are frequent droughts. In the 1960s there were six successive drought years. As the sketch map shows, among the fields are tanks – artificial reservoirs or ponds – which are fed from the streams flowing off the hill. In these, fish are kept, to be shared out among the villagers. The tanks are also used for small-scale irrigation and for watering livestock. They are a common-property resource, overseen by the elected village council.

The village is relatively homogeneous in terms of occupations and does not have wide social divisions. Most people are small farmers, and the dominant caste is Khandait, which comes relatively low down in the caste hierarchy. There are ten Scheduled Caste families, who live in a small settlement outside the village centre (see Figure 3), and who are landless. Land holdings for the rest vary between 0.1 ha and 4.0 ha. The average size is about 0.4 ha, which means that most farmers are 'sub-marginal'. (Just over one hectare would provide a family with sufficient subsistence and cash income to be considered secure.) To supplement their food and income from their fields, most people keep a few cows and goats, and grow vegetables in their back gardens. A typical family with land works less than one hectare of fields, has two cows, between two and four goats, and a few chickens. They also depend on the forest for small timber, fuelwood, fodder, and herbs. Some of the poorest families, who live mainly by labouring, are forced to seek work outside the village as seasonal migrant labourers on distant sugar plantations.

Although the village is now very cohesive, it was not always so. Dr Narayan Hazari, who grew up in Nayagarh and Kesharpur, describes Kesharpur in the 1950s and 1960s as being 'bedevilled by factionalism':

> There was a lack of collective ethics and an organised civil sense. Ignorance, suspicion, and fear ruled the village. There was violence in the air. There was no love. Factions ran to the police and the court to settle scores against the enemy. All the time, the energy and the money of the people was spent to destroy the opponent. There was economic drain from the village. The poor became poorer.[7]

The state of the forests

As late as the 1950s, the forests on Binjhagiri hill were virtually undisturbed, with, according to Banchhanidhi Pradhan of Gamein, *'trees of six or seven feet girth'*. There was also an abundance of large and small game, including bears and tigers. But after Independence people started to feel that they could do what they liked, since the land was now 'theirs'. Sudarshan Malla of Beguniapatana told us, *'People perceived Independence as liberty, without any commitment or responsibility.'* As a result, the destruction of the forest rapidly gathered pace. By the late 1960s, Binjhagiri was completely denuded, gullies gashed its sides, and rock waste was smeared over the fields below, making them infertile. It was the same for Malatigiri. As a result, the people of Kesharpur and neighbouring villages, such as Gamein and Beguniapatana, faced extreme hardships, since much of their non-agricultural subsistence needs, for fuelwood in particular,[8] were met by the forests. These shortages led some to desperate measures. An incident in 1978 is described vividly by Narayan and Subas Hazari:

> We had in our joint family a man named Balia who used to help us with our domestic chores. When the joint family broke up, Balia, a bachelor who had worked for us throughout his life, was thrown out of employment. It was like throwing him to the wolves. Balia, in his late forties, found shelter with his younger brother. He had no source of sustenance. He became prematurely old and suddenly died. His brother could not arrange wood for the funeral [because of the deforestation]. Balia was thrown on the sands of the river Kusumi which flows beside our village. Next morning it was found that Balia had been eaten up by jackals and his limbs were strewn everywhere.

Before this shocking event there had been some attempts, particularly by young people in Kesharpur (and in one or two other villages near the hill), to re-afforest and protect Binjhagiri. But the Hazari brothers describe this as having been done 'half-heartedly', in that it did not have wide community commitment and involvement. This was hardly surprising, given the factionalism that had so bedevilled Kesharpur. The fundamental problem was that, since Independence in 1947, people's relationship with the forests had completely changed from one which was sustainable and constructive to one which was exploitative and destructive, a transformation which is considered more fully in Chapter 3. The Hazaris attribute this to a number of factors, and they describe the wider consequences:

People's fear of the government lessened. Population grew. The government saw the forest as a source of commercial profit. People did not understand the utility of maintaining ecological balance. There was an attack on the environment. Trees were felled and forests were denuded. Hunters had a free day to play with the life of the fauna. The green Buddhagiri hill[9] on which our eyes feasted slowly and steadily became bald and brown. Our hearts filled with anguish to see scars on the body of Mother earth. There was fast soil erosion, and deep ravines came up. Agricultural fields were covered with soil and pebbles from the hill. There was a shortage of fuel. Within hardly a generation had come a sea-change for the worse.

That change for the worse also included a decrease in rainfall, an increase in droughts, and what many described to us as 'excessive hot air in the summer months', blowing off the desiccated hill sides. Such was the damage to agricultural productivity that many people had to migrate considerable distances in search of work. According to Dambarudhara Majhi of Binjhagiri village, some even went as far as Burma, never to return. It was in this context that Narayan Hazari started writing letters to 'my villagers' from his place of work (Utkal University, Bhubaneswar, where he was Reader in Public Administration), about the state of the environment and the importance of environmental preservation. He also went back to the village in his vacations. There, with members of his family and other local people, notably Joginath Sahoo, the headmaster of the local Middle Education School, and Udayanath Khatei, a much-respected Kesharpur farmer and village leader, he campaigned with the villagers to work on environmental protection. 'But', he writes sadly, 'everything fell on deaf ears. It did not click.'

The village is stirred to action

Following the dreadful fate of the body of Balia, a breakthrough came in 1978 when the National Service Scheme (NSS – a student movement established to do community-service work) held a summer camp for afforestation in Kesharpur. Working with people from the village, the NSS volunteers planted saplings on the hill. On the final day of the camp, there was a meeting attended by Pratap Patnaik, the local Divisional Forestry Officer (DFO), who urged the villagers to preserve, protect, and regenerate their forest. He also pointed out to them the incompatibility of planting and protecting young trees and keeping goats. (Patnaik was to become a valued friend of the movement in its early years, a rare example of a Forestry Officer who was on the side of the people.) Narayan Hazari writes of this event: 'It clicked. What I could not do for several years, this meeting miraculously achieved.'

In the months following, the villagers organised more camps, with NSS involvement. They also re-instituted an ancient system of forest protection, known as *thengapalli*, or 'stick rota'. Every evening two sticks were left at the doors of two adjacent households in the village. This meant that someone from each of those households was responsible the following day for patrolling on the hill. As they walked they would carry the *thenga*, the stick, as a symbol of authority and responsibility. With this they would ensure that no one entered the forest with an axe and no one took anything from the forest other than what was allowed (for example, dried sticks for firewood). When at the end of the day the *thenga* was left outside a neighbour's door, authority and responsibility were passed to that person. Every family in the village agreed to participate. Although this was done on a 'voluntary' basis, the sanction and force of the village council lay behind it.

The villagers took another decision which was momentous in its consequences, particularly for the poorer members of the community. Accepting the advice of Pratap Patnaik, the village council decided to ban all goats from the village for a number of years. As Narayan Hazari comments, 'This was not a small sacrifice for the poor villagers, considering the economic value of the goats.' However, in order to help the families most affected by this decision, that is the landless and most marginal farmers, loans were made from the village fund to help them to rent land from the larger farmers, and to rent ploughs and oxen. Through this measure the most vulnerable families could provide themselves with a subsistence income to match that derived from their goats. In time the village council also allowed people to collect broken-off branches, and dried leaves and twigs for fuelwood. They also allowed the cutting of unwanted plants, such as the shrub *pokasunga* (*Eupatorium* sp.), which grows profusely.

Within a very short time the villagers were able to see the benefits of protecting and replanting the hill. Grasses, which had previously been nibbled down to their roots by goats, grew longer and provided grazing for cattle. There was a reduction in the amount of soil erosion on the hill and in the deposition of rock waste on to the fields below. The gullies that had slashed the hillside began to grass over and heal. The smaller gullies were plugged to check the speed of water flowing down them. Trees started to sprout from roots and seeds. There was particularly prolific growth in the number of *babul*[10] trees, which provide wood for agricultural implements, fencing, and fuel. The fruits of the *babul* are eaten by cattle but not digested, so that they pass out in their faeces. *Babul* saplings started to grow wherever the cattle grazed. There were two places in particular where this growth was prolific. One was in a designated area behind the village where a patch of 'cow-dung forest', as they called it, quickly sprouted. The other was down at the riverside, where cattle came to drink. Here growth not only helped to protect the banks from erosion, but also provided a harvest of timber and fuelwood when the trees were mature. In a typically equitable arrangement, the river-bank harvest was divided 50:50 between the landholder of the adjacent fields and the rest of the village. Now, in Kesharpur, Bhobani Dora told us, *'We have plenty of babul trees. When they are auctioned, we buy them for fuel.'*

Once the forest was mature enough, wildlife returned, and particularly birds, including hundreds of cranes, which are important for keeping agricultural pests under control. Their presence enabled organic farmers to avoid the temptation to use pesticides.

However, there were problems. As the forest regenerated on Binjhagiri, protected by the people of Kesharpur, the threat of pilferage by nearby villages not involved in forest protection increased. It became necessary, therefore, to spread the message of environmentalism beyond the village to others, some of whom had independently started their own programmes of protection and regeneration.[11] This was done through staging rallies, acting out dramas, singing songs, chanting slogans, and conducting *padayatras* (campaigning marches). The work was tireless, focused, and determined. The philosophy behind it was Gandhian, and the tools were Gandhian too. Numerous stories are told by the Hazaris of this work. Here is one of them:

> In 1978 July 5000 saplings were planted [on Malati Hill] under the auspices of the Forest Department and the children of Manapur, through the untiring efforts of Sri Joginath Sahoo ... [However] this plantation was deemed a threat to the livelihood of the people engaged in quarrying and their contractor. In October 1978 when it was raining at night, some of them mercilessly uprooted the saplings. In the morning when Joginath learnt this, he ran to the spot and found that his dream was in a shambles ... Subsequently [he] along with his friends ran several times to the Revenue authorities in Nayagarh to persuade them to stop the auction of the hill for quarrying, which destroyed the eco-system. He was ridiculed, and his motives were suspected ... His repeated knocks (he went 13 times) ... fell on deaf ears ... He was also threatened to be murdered by the Contractor and his men. Finally with the intervention of the CCF [Chief Conservator of Forests] ... and many others, he succeeded substantially and the major part of the hill was made available for afforestation.

Action spreads through inspirational leadership

Over the four years from the first major breakthrough in 1978, reluctant villagers were slowly won over by the patient and inspirational work of the core activists, led by Joginath Sahoo. The importance of the role of *Shramik*[12] Jogi cannot be over-emphasised. He is a man of great vision and humility, for whom environmental protection became his *raison d'être*. Though not widely travelled, he is widely read. When asked recently what in particular, from all his reading, had spurred him to devote his life to environmentalism, he spoke of the impact of reading the poem 'The Deserted Village' by the eighteenth-century English writer, Oliver Goldsmith. Although 'sweet Auburn' of the poem was deserted for different reasons, Jogi saw that Kesharpur could become similarly abandoned, for the simple reason that the environment had become so depleted and degraded that it could no longer sustain the livelihoods of the villagers who depended on it. In other words, the very survival of Kesharpur as a community was threatened and in the balance.

He felt this for several reasons. There was at the least the *perception* that rainfall had diminished – certainly it seemed that the incidence of drought years had increased. Undoubtedly springs had dried up (it was reckoned that the water table had fallen by four metres), and the hillsides were bare and slashed with gullies. Tanks (reservoirs) were becoming choked with silt, and fields were covered with rock waste. Crop yields were falling, and grazing had become precarious because there was little grass for pasture and even less fodder for gathering. Fuelwood had become so scarce that people resorted to digging up roots of trees that had been left in the ground; and many people were cooking only one meal a day because of the fuel shortage. And wildlife had disappeared along with the forests. *'There was not even a rabbit on the hill,'* Neema Pradhan, the village health worker of Gamein, told us.

Jogi was essentially telling people that unless they did something to conserve their environment, the village was doomed, and the children whom he was teaching would have no future there. Kesharpur would become another 'deserted village'. What he offered them, as an alternative to this predicted doom, was a vision of a village which could live in harmony and balance with its environment, in which the villagers took from it only what it would *sustainably* yield.

In devoting his life to the movement, Jogi has made huge personal sacrifices, the greatest of which has been his family. In the way of extended families in India, his children have been brought up by his brother and sister-in-law, to allow Jogi and his wife to devote all their time outside school hours voluntarily to the movement.

The birth of Friends of Trees and Living Beings

Throughout 1982, Jogi and another key person from Kesharpur, Udayanath Khatei (affectionately known as *Bapa* – 'Father'), together with a small group of teachers, held a number of meetings in the villages around the hill. 'We moved between the villages,' Bapa told us, 'persuading people of the need for environmental protection.' Then in October 1982 there was a large informal meeting outside the youth club in Kesharpur. To this meeting came at least five people from each of the 22 villages. Also present were some other key people, including Professor Radhamohan and Dr Narayan Hazari from Utkal University in Bhubaneswar, and Bhagaban Prakash, NSS Coordinator at the University. These played an invaluable role in advising and facilitating the momentous meeting. 'We didn't really know what we were doing', Bapa told us. 'We were just feeling our way. The outsiders were very important in moving us forwards.'

The meeting had two key outcomes. First was the decision to form an organisation with a set of objectives and a name. Initially it was proposed that they should call themselves 'Friends of Trees', until one of the visitors pointed out that there was an organisation in Bombay with that name. So they agreed on 'Friends of Trees and Living Beings' – *Bruksha O' Jeevar Bandhu Parishad* (BOJBP). Second was the decision to form a committee under the leadership of Jogi and Bapa. In the weeks that followed, two other key meetings were held in Nagamundali and Gamein; out of all these meetings, plans of work and a set of rules were drawn up and objectives agreed. Among the key resolutions which emerged were the following:

1. The *Anabadi* forests or fields at the foot of the hills should not be given to anyone on lease. (This was land within the village boundary that was managed by the Revenue Department. It was necessary to protect it from encroachment, in order to allow it to regenerate and recover.)
2. Officially licensed quarrying in the hills should be stopped.
3. Villagers should be involved in plantation programmes, so that they would become stakeholders in their preservation.
4. Goats and sheep, considered deadly enemies of growing plants, should be sold and should not be re-introduced for at least ten years.[13]
5. To meet cooking needs, fuelwood would be provided by the Forest Corporation[14] at a reasonable price. Alternative sources of energy for cooking would be popularised, and fuel-efficient stoves would be promoted, to prevent pressure on the forests.

6. People should be encouraged to adopt family-planning methods to reduce pressures in the future on forest resources. (*'Increase the number of trees; decrease the number of children'* was the slogan.)

Who was responsible for the formation of BOJBP? The Hazaris suggest six 'agencies'. First they cite 'the masses': the villagers themselves.

People power

Although in the 1950s the people of Kesharpur had been at each other's throats, over time they learned to work together, and through 'voluntary, co-operative community endeavour' they built two schools and renovated the village tanks (reservoirs). Similar action was also noted in other villages. The Hazaris refer, for example, to the ten villages in the Kalikaprasad *panchayat*[15] which donated half a year's sugar quota (from the Public Distribution System) to raise Rs. 90,000 to build a dispensary. As the Hazaris observe, considering the 'isolation and narrowness of the village situation, the working together of ten villages for the setting up of the dispensary was most remarkable'. They also point out the importance of the role played by the masses in keeping the leaders going 'when morale was sagging', and they comment that giving inspiration was never a one-way process:

> The masses go beyond the expectations of the social leaders, give fillip to their sagging spirits, inspire them and give a call [to them] to rise to the occasion. If the social leaders do not come forward to meet this challenge, the programme's future is doomed. The role of the masses should not be underestimated.

Implicit in the relationship described here is what Pradeep Pattanaik, BOJBP Field Organiser, calls the 'emotional bondage between people and BOJBP'. It was because this bond was so strong, and because there was a strong culture of voluntary labour in these villages, that so many people could be readily mobilised in the activities of the movement. Banchhanidhi Pradhan of Awasthapada told us, *'The leaders made the people sensitised in such a manner that if a meeting is convened 100 people could easily gather.'* The closeness of the relationship between the villagers and their leaders was to be a key feature in the movement's successes for at least ten years, which makes it all the more painful to report, as we do in Chapter 6, that it has recently broken down.

Inspiring leaders

The second group of key players are the 'institutional' village leaders, represented in Kesharpur by a man like Udayanath Khatei (*Bapa*), who was for many years President of BOJBP. Bapa's authority and respect come not through his wealth (he is a marginal farmer), nor his caste or title, but through his service, sacrifice, and hard work in the cause of the movement, and through his identification with the community. Because he was one of them, he played a crucial role in motivating fellow villagers. Biswanath Basantia, a former Secretary of the movement, told us that it was because of Bapa's position as respected village leader that he could so easily harness 'the villagers and the village common resources for strengthening the movement'. Other villages had their own *Bapa* figures.

Social activists

Third, there are the social leaders – perhaps better called leading social activists – of whom there are two types. The first group, represented by Joginath Sahoo, live within the community and take on the mantle of leadership. The second group is represented by Dr Narayan Hazari himself, who, while keeping in touch with the grassroots, stays at a distance from the scene of the action, but close to the seat of power. In Hazari's case he lived and worked in Bhubaneswar, where he had access to research organisations, the NSS, other development agencies, and government officials. Others of his kind, though less intimately involved in the movement, included Professor Radhamohan, who has always provided wise counsel.

Several people commented to us on the twin roles of Dr Hazari and Joginath Sahoo. While acknowledging the huge significance of Jogi as an inspiring organiser – a *'selfless, hard-working, down to earth person'*, in the words of Pradeep Pattanaik of Gambhardihi – many people interviewed for this book paid a special tribute to Dr Hazari *'for taking the lead in educating people for environmental protection'*, especially in those crucial early years. Thus, Manoj Hazari, a youth activist from Kesharpur, commented:

> Professor Hazari was the real spirit behind the ... movement in the area. Frequently he was coming to the village in the holidays ... to unite youth and children. Through different games and entertainment programmes, he used to mobilise the children for welfare work.

To this Susanta Jena, BOJBP's first paid worker, added, *'He awakened the organisation by writing letters frequently and instructing the leaders to do this and that.'* Whichever one of the two leaders played the most significant role is pointless to argue: they both had their parts to play at different times.

Within the group of social leaders, Dr Hazari points out the special part played by school teachers (of whom, of course, Jogi was one):

In the rural community they are the only people having leisure. They can work for the community on holidays and this does not affect their pay packet, whereas the position is very different in [the] case of the peasants and the landless farm workers. In this movement school teachers were the most vital component of social leadership ... Today almost every Indian village has a school. If the school teachers take up the responsibility of mobilising the masses for community development, the face of the country will change for the better.

In addition to their potential role in directly 'mobilising the masses' in their spare time, teachers also play a critical role in educating children through the formal curriculum and extra-curricular activities. The involvement of children in the movement has been extremely significant, for not only can they take part in forest regeneration through plantation work outside school hours, but they can also be, in the words of Gadadhara Bhatta, Headmaster of the Middle School in Kalikaprasad, 'the main carriers of the message' to their families and wider communities. Children and teachers can benefit through learning together. To quote Gopal Pattanaik, a former teacher in Awasthapada: *'Not only [did] children educate themselves, but also we teachers learned many things from the process. [In addition] inter-school relationships were strengthened because of the regular interactions, competitions, and collective interventions.'*

Furthermore, once sensitised, school children could even be used in the front line of forest-protection work. Thus, Madhusudan Dora, a Class 7 pupil in the Middle School in Kesharpur, described to us a memorable occasion:

> .... people from Sanagarada and Badagarada were cutting from our forest. When we [children of Kesharpur] opposed them, they threatened us by showing us knife and axe. We ran into the village and informed the elders, and 40 to 50 people from our village went to the forest and caught hold of the woodcutters.

Jogi himself told us of an occasion when he discovered that some herders were letting their cattle graze in an area of plantation. His response was to get 150 school children to go to the village where the herders lived, to sleep on the road in front of their houses. Seeing this little army of young protesters, the herders *'trembled and could not say anything. All of them promised not to graze their cattle in the plantations and forest.'*

When talking to us about leadership, Jogi was quick to point out that in addition to the school teachers there were others in the villages without whom they would never have succeeded, who provided second-line leadership. They include Bhikhyakari Hazari, Biswanath Basantia, Trinath Jena, Gatikrushna Bhatta, and Sankar Bhatta, and many others.

Voluntary agencies

The fourth group of actors is the voluntary agencies, represented in BOJBP's early years by the NSS units from Utkal University, and Nayagarh and Sarankul Colleges. Not only did they contribute labour in tree planting, but with their youthful enthusiasm and commitment they led by example. They were especially important in providing role models to school children. *'They encouraged us through cultural programmes'*, Manoj Hazari of Kesharpur told us. *'We co-operated with them in planting trees on the hill.'* There is no doubt that, without them, progress towards developing the movement would have been much slower. They were particularly important for their involvement in several workshops that were held in 1983 following the birth of the movement. In addition to the students themselves, three college lecturers from Nayagarh individually gave significant support to the movement: Parikhit Nanda, Nakula Rath, and Saroj Pattanaik.

Government agencies

Fifth are various government agencies, most especially the Forest, Environment, and Revenue Departments. Relationships were not always good between the movement and these departments, but at various times they all played decisive roles, in giving moral support, technical advice, and encouragement. Over the years many officials visited Kesharpur and were amazed and genuinely inspired by what they saw. In return they spoke encouragingly to people involved in the work. After visiting, one of them, the Divisional Revenue Collector, wrote: *'If you want to see a place where development has come through people's participation, then come to this place.'*

One of the most valued friends of the movement was Pratap Patnaik, for many years the Divisional Forestry Officer (DFO), already mentioned in connection with his stricture on goat-keeping to the villagers of Kesharpur. Another was the Chief Conservator of Forests (CCF), Sachidananda Das, and his successor, Padarbinda Mohapatra, who visited the area once more.

It must be admitted, however, that at the local level the movement was not supported by many of the foresters. Frequently the Range Officers were obstructive, manipulative, and downright incompetent. To illustrate the problems from this quarter, Narayan Hazari tells this story:

> The Forest Department had sanctioned Rs. 100,000 for afforestation in the village of Gamein in July 1981. Two watchmen from the village were paid by the Forest Department to keep watch on the plantation. After five months the Range Officer stopped paying them. They complained ... but it fell on deaf ears. So they discontinued their watch. Cattle had a free time to destroy. Along with

students and five teachers from the college, I visited the village after seven months of ... carnage. The Range Officer, who had never bothered to visit the site, got wind of our visit and came from a distance of 18 km and followed us. Perhaps he was apprehensive that we would discover his lapse and convey it to the Chief Conservator of Forests, to whom we had access. When asked why he did not visit the area, he offered two reasons: (a) that he had to remain busy with catching timber-theft cases in villages which plundered the forest and (b) that it was not possible for him to come to the site [on his] motor cycle because of [the] bad road. His explanations left none of the listeners convinced.

Politicians

Finally, Narayan Hazari identified a sixth group, which could have played a significant role in the early years of the movement but failed to so: politicians. Of these he writes, somewhat tersely:

> Political leaders, incidentally, did not play any significant role in the movement. In fact, a local politician and one of the most powerful political functionaries in the State played a negative role. But it is possible [in theory] to visualise a situation where political functionaries could be very useful agents of change and progress.

Summary

The struggle was constant, and the need for vigilance never decreased, as subsequent chapters will tell. However, what is evident from the story of these formative years is the vision and commitment of the leaders, both institutional and social. Narayan Hazari was a key figure, providing the embryonic movement with its early inspiration, and later with intellectual support and access to other sympathetic players in Bhubaneswar. Udayanath Khatei was a respected local figure whose word would always be heeded by people in the villages. And Joginath Sahoo, supported by teaching colleagues and others, was the organiser, activist, and leader, capable of galvanising whole communities to action. To these must be added all the volunteers from the various NSS units, who turned up to plant trees, to run workshops and camps, and to inspire the villagers with their commitment and enthusiasm. Finally there were the people themselves – adults, youths, and children – who, little by little, over many years, were persuaded to opt for a vision of hope instead of a future of doom, and to become part of a remarkable movement.

The wider picture 2

Deforestation seems to have come to symbolise the situation of the overexploitation of natural resources. Unfortunately, as an old Thai saying goes, 'experience is a comb which nature gives to a man after he is bald'. Today, even in the face of mounting disasters resulting from the assault on nature, we are fast reaching the 'baldness' state.
(Walter Fernandes, Geeta Menon, and Philip Viegas[1])

India's forest wealth

India is one among the twelve countries that collectively account for 60–70 per cent of the world's biodiversity. India's forests, grasslands, wetlands, and marine eco-systems are the homes for no fewer than 46,000 plant species and 81,000 animal species, and they contribute 8 per cent to the known sum-total of global biological diversity.[2]

Of all India's richly varied eco-systems, it is the forests that are best known for their immense complexity and diversity. There are 16 important forest types, categorised into five major groups. These are Tropical forests, Montane sub-tropical forests, Montane temperate forests, Sub-alpine forests, and Alpine forests. Of these, Tropical forests are the most extensive. Partly reflecting this dominance is the fact that 85 per cent of the country's forests are broad-leaved hardwood species, whereas only 6 per cent are coniferous, softwood forests.[3] (The remaining 9 per cent are bamboo forests.)

Though now heavily degraded, India's forests have from time immemorial protected India's environment by maintaining its ecological balance, protecting and enriching its soils, modifying its climatic extremes, and regulating its hydrological cycle. They have also provided a habitat for tens of thousands of animal species, and for tens of millions of people, who have developed their societies, evolved their customs, and woven their myths around and within them.

Legal classifications
At the time of Independence in 1947, India's forested area was estimated to be 40 million hectares (mha). By 1950–51, because of the absorption of the princely States and *zamindaries*[4] into the Union, this had increased

significantly to 68 mha. Now, for reasons given later, the official (recorded) forested area has grown yet further to 76.52 mha. However, as will be seen, satellite-imaging casts doubt on the accuracy of this figure. Of the total official figure, 54 and 29 per cent are designated respectively as 'Reserved Forests' and 'Protected Forests'. (The rest are 'Unclassed Forests'. In some States there are also 'Village Forests'.)

Reserved Forests are those managed exclusively by the government through the Central and State Forest Departments, both for revenue raising and for environmental protection. Normally no rights of usage (usufructuary rights) for local people can be entertained in Reserved Forests. Protected Forests, on the other hand, are for meeting the needs of the people at the local level, who in return for the payment of taxes and royalties have certain rights of access and concessions in these areas. Protected Forests are divided into two categories: 'Demarcated Protected Forests', meaning that they have been delineated, and 'Undemarcated Protected Forests', which have not. As far as Orissa is concerned, while large areas of forest – Reserved, in particular – have been transferred from the Revenue Department to the Forest Department, huge swathes of Protected Forests have still not been properly identified and brought under its control. They remain, therefore, Undemarcated and under the control, as we shall see, of the Revenue Department. The *Khesra* forests of Binjhagiri and Malatigiri are in this category.

The origins of India's Reserved Forests go back to the 1880s, when the scientific management of forests started. The government at the time identified large areas of forests, registering them as Reserved and keeping them under strict control. This process of forest reservation is still continuing, in spite of the fact that large areas of so-called 'forest', even newly reserved forests, are highly degraded, and other areas have been diverted for non-forestry purposes. This explains why the official figures show that the forested area of the country is still increasing, when in reality much of it is no longer adequately tree-covered, or even tree-covered at all. This discrepancy is exposed in the reports of the Forest Survey of India (FSI), a government institution which, using satellite imaging, estimates the extent of forest cover in the whole country every two years. Table 1 reveals that not only are the FSI figures substantially lower than the official figures, but the state of the forests, as expressed in terms of crown density, shows them to be in an even worse condition.

Half of the forests in Orissa are under the control of the Forest Department, and the other half, comprising 'forest-land' but very little actual forest, are under the administration of the State Revenue Department. This arrangement is peculiar to Orissa and has come about in the following way. After the

Table 1: Forest cover in India and Orissa according to official and satellite imaging data

Forests	Recorded/official, 1997		FSI from satellite images, 1997[5]	
Area	Total mha.	Per cent	Total mha.	Per cent
Forests throughout India	76.52	23	63.33	19
Forests with dense crown throughout India			36.72	11
Forests in Orissa	5.72	37	4.69	30
Forests with dense crown in Orissa			2.63	17

(Source: Forest Survey of India and records of the Forest Department)

post-Independence merger of the princely States and *zamindaries* with the Indian Federation, their forests were vested in the government. But the Orissa Forest Department at that time was not in a position to manage these new responsibilities. Therefore, the forests of the ex-States and ex-*zamindaries* were treated as Protected Forests (within which people who were paying taxes continued to exercise their rights and concessions). But, unlike the Reserved Forests, they were never actually *transferred* to the Forest Department, as they could or should have been, through existing legal frameworks. Therefore, because no forest settlement was made for the Protected Forests, the land is still owned by the State Revenue Department, but the State Forest Department is actually technically managing the forests still standing on it.[6] Not surprisingly, there are often conflicts between the two departments relating to Protected Forests.

Economic and social functions of forests

Although 95 per cent of India's forests are under government ownership, they are also the habitat of more than 50 million tribal peoples, who rely on forest resources for their subsistence. Among these are artisans who require raw materials for their crafts. In 1981 these numbered more than 3 million,[7] consisting of bamboo and cane workers, potters and wood-workers supplying products to millions more through local markets. There are also forest farmers and their families who cultivate forest clearings, many on a shifting basis, and gather supplementary foods and other non-timber forest produce (NTFP) from the forests. With their families, these number many millions. (The next chapter considers their conditions in detail.) In addition, between 250 and 350 million more are dependent on forests to some lesser but still significant degree, for fuelwood, fodder, herbs, and small timber. And there are the hundreds of millions living in towns and cities who cook with fuelwood cut from forests somewhere. In addition there are those millions working for contractors in logging and in the pulp, paper, and timber industries. With all these hundreds of millions directly or indirectly dependent on India's forests, it is hardly surprising that the forests are so endangered. Nor is it surprising that those who are most dependent on them – the actual forest-dwellers themselves – have so little left on which to live. While forest-dependent people minimally require on average 0.5 ha of forest per capita, the actual availability is 0.1 ha.[8]

In addition to the material importance of India's forests, they also have two other important functions for those living in close association with them. First they provide a space for cultural activities and entertainment, bringing people together in a shared experience: the forest is their stage, their theatre. Second, as we have observed within the BOJBP 'mother area', forests play a vital role in generating unity and solidarity within communities, which comes through the experience of working together in the communal management of the forests. In this way forests have become crucial in village development.

Evolution of the legal framework and the alienation of the forest dwellers

The national picture

Traditionally India's forests were managed to meet the needs of local communities, that is communities living in or close to forests, and were governed by customary rules and regulations which had evolved over the centuries. However, this traditional framework of forest use and governance was over-ridden in the second half of the nineteenth century, when the British saw the commercial value of India's forests and imposed control over them in the name of 'scientific forest management'. The first move in this direction came in 1855, when Lord Dalhousie, India's first Governor General, issued a memorandum, entitled 'Charter of Indian Forests'. In 1865, the foundation of the Forest Department was laid with the promulgation of the first Indian Forest Act, which empowered the government to declare any land covered with trees or brushwood as 'forest' and gave it the right to control it. Thus, at a stroke, forests that had hitherto been considered as common-property resources came under the authority and control of the government. The Indian Forest Act 1878 classified forests as Reserved, Protected, and Village Forests, and defined usufruct rights accordingly. In 1894 a Forest Policy was drawn up, emphasising the need for State control over forests in order to augment revenues. Later came the 1927 Forest Act, which reiterated and strengthened the provisions of the 1878 Act. This (1927) Act, with some amendments, continues to provide the legal framework today, although later acts have been passed to tighten it.

In 1952, five years after Independence, a new National Forest Policy was framed. This recommended that, in order to preserve ecological balance and security, at least one-third of the country's area should be kept under forest cover, a recommendation that has been totally ignored. Under this policy, shifting cultivation was to be discouraged and forest grazing regulated. Although the policy emphasised the ecological and social aspects of forestry (by giving secondary importance to commerce, industry, and revenue raising), actual government practice went against it. As a result, large areas of forests in ecologically sensitive areas have been destroyed to make way for big projects – power, mining, irrigation, and industrial ventures, and infrastructure such as roads and railways. Furthermore, through the contract lumbering system, huge areas of forest have been clear-felled to raise revenue for the State.

As if this were not enough to perpetuate the assault on India's forests, the 1976 National Commission on Agriculture strongly advocated further commercialisation and emphasised the need to plant 'production' forests in order to maximise forest yields. On the basis of the Commission's recommendations, a new Forest Bill was drafted in 1980, which many described at the time as 'draconian' for its 'anti-people' provisions, but which, because of strong opposition by social and environmental activists, was never tabled. In its place came the relatively pro-people, pro-forest Forest (Conservation) Act, 1980, which, as its name suggests, emphasised forest conservation and put a brake on the diversion of forest lands for non-forestry purposes.

In 1988 a further Forest Act strengthened the 1980 Act, forbidding State governments to lease out forest lands without the prior sanction of the central government. This was a turning point in the history of forest management in India. Alhough it established the centralisation of power for forest conservation, it was widely accepted as necessary to save India's rapidly degrading forests from rapacious exploitation by industrial and commercial corporations. Furthermore, under those circumstances where the diversion of forestland for development purposes could not be avoided, compensatory afforestation (about which more is said later) became mandatory.

The National Forest Policy, 1988, was another historic move by the central government. This gave precedence to environmental and ecological concerns over economic benefits. It re-stated the national goal to keep one third of

India's land under forest cover, though in reality that threshold had long been passed. The policy also prescribed full protection of the rights of tribal and other forest-dwelling people over the forests. Indeed, meeting the needs of local populations dependent on the forest eco-system was held to be the first charge on the forest. This meant that forests could not be exploited to meet the raw-material demands of industry, nor earn revenue for the government *at the cost of local populations*. For the first time, forests were conceived as local resources, to be managed in such a way as to maintain environmental stability. This policy was, therefore, a significant departure from long-standing forest-management practices whose emphasis had been on commercial exploitation and revenue raising.

However, although the policy resolution was adopted, no steps have been taken by the central government to implement it through the enactment of new forest laws. So in fact the State governments are still pursuing the same forest practices under which forest dwellers' interests are not properly protected, indeed are ignored and trampled on. This is an issue to which we shall return in the next chapter.

Forest management in Orissa

Orissa's forests are managed by the Orissa Forest Department under the mainframe law of the 1972 Orissa Forest Act. Various other acts and rules have since been passed to facilitate the management of the State's forest and forest produce.

About 45 per cent of Orissa's forests have been legally classified as Reserved Forests, under the control of the State Forest Department. All but a very small proportion of the rest are Demarcated and Undemarcated Protected Forests, still under the control of the State Revenue Department, for reasons that have already been explained. (The tiny fraction that is not controlled consists of Village Forests.) No efforts have been made to legally register the Protected Forests, nor to develop operational rules to manage them, which could and should have been done under the provisions of the 1972 Orissa Forest Act. The reasons for this almost certainly relate to the unresolved conflict between the State Forest and Revenue Departments that has been noted. Until the 1980 Forest (Conservation) Act, Orissa's Protected Forests were under constant assault through encroachments and deforestation. In fact, by the time the Act was passed, most of Orissa's Protected Forests had already been diverted to other uses, without even the knowledge of the Forest Department, let alone its permission.

This does not mean that the Orissa Forest Department has behaved benignly towards forest people, even in Protected Forests where communities have *bona fide* usufruct rights. Far from it: it has imposed a whole variety of restrictions upon them. For example, as explained more fully in Chapter 3, most NTFP has been nationalised, so that the State can control its procurement and sale. This move has resulted in thousands of people being drawn into highly exploitative labouring work for monopolistic private lessees and departmental agencies involved in the marketing of NTFP. There is still no mechanism for ensuring forest protection in the interests of vulnerable forest-dependent people.

Vanishing frontiers: deforestation and the degradation of forest resources

> Deforestation is the inevitable result of the current social and economic policies being carried out in the name of development.[9]

The wholesale destruction of India's forests encouraged a devil-may-care attitude towards the forest among rural communities. From colonial times onwards, bereft of their traditional rights, people regarded forests as an openaccess resource and so contributed further to their destruction through encroachment, that is the illegal diversion of forestland for private purposes, such as housing and farming, and for communal purposes, such as schools, temples, and recreation. Now, according to official records, approximately 7 mha of Indian forestlands are under encroachment – an astonishing figure.

Between 1950 and 1980, before the implementation of the 1980 Forest (Conservation) Act, 4.3 mha of forestland had been diverted for various non-forestry purposes, of which 2.6 mha went to agriculture and the rest to development projects. (These figures do not include encroachment.) Following the 1980 Act, the diversion of forests for non-forestry purposes was regulated and slowed considerably. There have also been powerful pressures on State governments by the Central Ministry of Environment and Forests not to appropriate forests for industrialisation and other development projects. Furthermore, 'compensatory afforestation' programmes (see below) have been initiated to compensate for the diversion of forests and forestland for non-forestry purposes. But in spite of stringent laws and tight policies, forests are still subject to enormous pressure from development projects. This has been all the greater with the opening up of the economy to private investment and the global market in 1991, which has made resource-rich areas of the country – and the people living in them – highly vulnerable to corporate interests.

The official records show that almost 0.3 mha of Orissa's forests were diverted for development projects between 1950 and 1998, but it is thought that this figure grossly underestimates the real state of affairs. Since major portions of the Protected Forests in Orissa, as elsewhere, have been encroached upon by rural populations without the knowledge or permission of the Forest Department, the actual amount will certainly be much higher. In 1983 the State Forest Department reckoned that encroachment stood at 0.074 mha. But since the amount of cultivated land in Orissa increased from the early 1950s to 1998 by 0.7 mha (from 5.6 to 6.3 mha^{10}), it can only be assumed that the bulk, if not all, of this large amount came from encroachment. The truth is that, as there is no current forest survey work, nobody knows the real state of Orissa's Protected Forests, which constitute more than 50 per cent of the State's forest resources. Although forests are there on paper, on the ground the reality is quite different.

Shifting cultivation

Shifting cultivation, which is considered in more detail in the following chapter, has been generally, but wrongly, perceived as a major cause of deforestation in India and elsewhere in the tropical world. Exact and reliable figures of the amount of land under this ancient form of farming are hard to find. But in India, in 1956, it was estimated that well over half a million hectares were under shifting cultivation, providing livelihoods for just over half a million families, with a total population of 2.64 million. By the early 1980s, the specially constituted Task Force on Shifting Cultivation noticed a sharp increase in the area of land under shifting cultivation. In spite of the fact that tens of thousands of shifting cultivators now practise settled agriculture, nevertheless approximately 4.35 mha are currently estimated to be 'affected' by shifting cultivation in India. Of this, Orissa accounts for more than 50 per cent. While it is true that large areas of forest have been affected by shifting cultivators, the cultivators themselves are hardly to be blamed for forest destruction. As we shall show in Chapter 3, shifting cultivation, as traditionally practised, is not of itself destructive. It is only when shifting cultivators become 'squeezed' by modern development schemes, by land alienation, by encroachments, and by Forest Department rules that they are forced into using unsustainably shorter periods of fallow. It is then that they – along with landless encroachers – do their damage. But such damage as they do is insignificant, compared with 'development collateral', and with biotic pressure, to which we now turn.

Ever-increasing biotic pressure

Biotic pressure is a major cause of deforestation in India. Indian forests constitute 2 per cent of the world's forest areas, but support 12 per cent of the world's human population and 14 per cent of world's livestock population. The sheer growth of India's population – by over 16 million per year – places huge strains on its forest eco-systems. One consequence of that pressure is the high incidence of forest fires, which now contribute massively to India's deforestation. In Orissa alone, the Forest Survey of India reckons that 94 per cent of the State's forests are prone to fires. Not only are rich forest resources severely damaged, but also their capacity to regenerate suffers enormously.

Then there is the pressure of domestic animals. India has a total of only 12 mha of pastureland and 'other' grazing land, little of which is properly managed in a way which could increase its productivity. As a result, it is estimated that about 90 million domestic animals graze in India's forests, whose carrying capacity is no more than 30 million.

Another crucial issue is the consumption of fuelwood, which nearly doubled between 1953 and 1987, that is from 86 million tonnes to 157 million tonnes. According to official figures the amount harvested in 1987 was only 58 million tonnes. This means quite simply that the rest was illegally procured from the forest, regardless of its carrying capacity.[11]

India's timber requirement in 1987 was estimated by the Forest Survey of India to be 27 million cubic metres (18 million tonnes), and at that time the official production figure was 12 million cubic metres (8 million tonnes). The difference between these two sets of figures can be accounted for only by illegal timber harvesting, that is smuggling, a subject to which we return below. In Orissa it has been estimated that by 2001 the requirement for timber will reach 0.367 million tonnes, with a massive shortfall of 0.267 million tonnes.[12] Similarly, Orissa's fuelwood requirement will be 14.128 million tonnes, with a highly alarming shortfall of 13.993 million tonnes.

Smuggling

Illegal felling, that is smuggling, is now a major cause of forest destruction throughout the land. In most areas it causes far more deforestation than legal felling. In the course of their inquiry in Orissa, to which we have referred, Fernandes, Menon, and Viegas[13] reported that the 'real use of forest material, particularly timber and bamboo, is probably four to five times greater than the official figures'. On condition of anonymity, one representative of a local paper mill told the researchers that for every truck of bamboo and timber taken out of the forest legally, his mill and others, as well as other forest-based industries, took three or four truckloads without a permit. The Fernandes inquiry also revealed the extent to which local politicians use their patronage to support this illegal trade, to which local forest officials turn a blind eye – in return, of course, for bribes. Further it was reported that when local people and honest foresters have tried to stop it, they have been met with threats, harassment, and even death at the hands of the police, Forest Department officials, and the smugglers themselves.

Forest regeneration and conservation

Plantation forestry

It is important to re-emphasise that in India, the 1980 Forest (Conservation) Act notwithstanding, forestry development through planned measures has long been conceived as protection, regeneration, plantation, and production *for the benefit of the State*. The social aspect of forest management is a relatively late concept which has yet to be reflected in practice in any of the planned programmes. Therefore, plantation for industrial and commercial ends is still the dominant forest-development activity. To this end, millions of rupees have been spent in raising plantations, rehabilitating degraded forests, and promoting farm forestry, and very little money has been invested in forest conservation.

And what is true nationally is also true for Orissa, where the major focus of forestry development has always been on forest regeneration through plantations. Between 1951 (the commencement of the First Five Year Plan[14]) and 1996-97, 824,776 ha have been planted under block plantation schemes. A further 297,493 ha have been afforested under schemes relating to the Rehabilitation of Degraded Forests. Furthermore, 30,830 million seedlings have been distributed to farmers and individuals for farm-forestry activities under the banner of the Social Forestry Programme, about which more is written later. Also, under 'avenue plantation', 11,802 km of Orissa's roads have been lined with trees. It should be pointed out that the bulk of the plantation work under these schemes involves mono-cropping, commonly with non-indigenous species – an initiative that creates its own problems, as we shall see.

Compensatory afforestation

The aim of the Forest (Conservation) Act of 1980 was to strike a balance between development and forest conservation. It stipulated that forests cannot be diverted for non-forestry purposes without the approval of the

Central Ministry of Environment and Forests. In those cases where there is a pressing need for conversion for development purposes, the same area of non-forestland has to be brought under compensatory afforestation. And if non-forestland is unavailable, the Act requires afforestation to be implemented in areas of degraded forest to the extent of double the area diverted for non-forestry purposes. Nationally between 1981 and 1997, 0.42 mha of forests have been diverted, and 0.5 mha of land have been earmarked for compensatory afforestation. However, only 54 per cent of that earmarked land has actually been brought under compensatory afforestation. It is the old story of strong legislation being enacted, but then to a great extent being ignored.

Furthermore, even where it is carried through, compensatory forestry does not replace like with like, since natural forests are almost always replaced with commercial mono-crop plantations of quick-growing species, destined to meet the demands of the pulp and paper industries. Such plantations have extremely poor survival rates, and certainly do not provide any compensation for the loss of the rich resource-base and habitat of millions of forest-dwelling peoples.

The rise and fall of social forestry

The theory

By the mid-1970s many had realised that the growth in wastelands and the unremitting pressure on India's remaining forests was so great that a serious policy reappraisal was needed. Such a reappraisal had to recognise the needs of local people, otherwise the forests could not be saved. Thus was born the concept of 'social forestry', whereby for domestic needs to be met, in particular the demand for fuelwood, fodder, and small timber, there had to be investment in the regeneration of village lands and private holdings to relieve the pressure on production forests. These lands were not forestlands *per se*, but a mixture of private farms, common lands, degraded village forests and other wastelands, which lay within relatively easy reach of settlements and which had been particularly badly affected by deforestation. The Indian Social Forestry Programme, launched in 1978, contained three elements:

- farm forestry, whereby private farmers were encouraged to grow trees on their own lands by the gift of free or subsidised seedlings;
- community woodlots, planted by the communities themselves on community lands;

- woodlots planted by Forest Departments on government lands, such as along the sides of roads and canals.

The practice

Between 1980 and 1990, social forestry was very actively promoted, with programmes in virtually all States. Notably successful were those in Punjab, Haryana, Gujarat, the western part of Uttar Pradesh, and West Bengal. However, closely scrutinised, the successes, in all these places except West Bengal, overwhelmingly relate to the farm-forestry element of the scheme, rather than the other two elements, that is community woodlots and government land plantations. Furthermore, right from the start and contrary to all intentions, the emphasis was not on species which would meet the subsistence needs of local people, but on fast-growing species, eucalyptus in particular, from which farmers could acquire considerable cash benefits by selling them for poles (for building), for pulpwood (for the paper and rayon industries), and for timber. And the farmers who were doing this were, of course, the larger farmers who were quick to seize the offer of free or highly subsidised seedlings (made possible through generous financial terms negotiated with the World Bank and other funding bodies). The figures showing the success of the farm-forestry element of the Social Forestry Programme are truly remarkable. In Uttar Pradesh, between 1979 and 1984, against the original target of 8 million seedlings, 350 million were planted. In Gujarat in one year, 1983–84, the target of 49 million was exceeded by a factor of four. In Punjab in one decade, more than 3 per cent of cultivated land was planted under eucalyptus.

Cumulatively all this planting had a significant impact on the net rate of Indian deforestation, which was much reduced. However, while this in itself was a good thing, in the terms in which it was conceived, the Social Forestry Programme was, in the words of the authors of a recent Oxfam report on Community Forest Management, 'a disaster in every possible way'.[15] Like so many well-intentioned development initiatives, it was subverted by the powerful, with the vigorous support of both the Forest Departments and the big international funders[16] who backed them. And all this to the detriment of those whom it was most intended to assist, those most dependent on the forests, whom we consider further in the next chapter. Furthermore it did not reduce pressures on Reserved Forests, which had been its primary purpose. In the words of Chambers et al.,[17] 'Wasteland development was the slogan, tree farming the reality.'

What followed was something rather different: Joint Forest Management (JFM), which we will now consider.

Joint Forest Management

The decade of the 1990s heralded a new era in the history of Indian forest management. A start was made in 1988 with the National Forest Policy, when forest-dependent communities were asked for the first time to become partners in forest management with State Forest Departments. Then in June 1990, the Central Ministry of Environment and Forests issued a circular with guidelines to all States and union territories for the promotion of community involvement in Joint Forest Management (JFM), which is seen as a partnership between communities and Forest Departments in the management of local forests.

By the end of 1999, following the circular of June 1990, 22 State governments had introduced their own JFM orders to encourage participatory forest management, and there are now 36,075 JFM committees in the country, protecting about 10.24 mha of degraded forest.

In August 1988, Orissa became the first State to involve local people officially in the protection of natural forests. The State Government allowed local communities to protect Reserved Forests surrounding their villages, in return for which they were granted certain rights of usufruct in those forests – even though they were Reserve Forests. Then in May 1990, similar provisions were extended to Protected Forests. In response to these initiatives, 6085 Village Forest Protection Committees (VFPCs) were established by the Forest Department to protect 1.42 mha of forests, of which more than 80 per cent were Reserved Forests.

Following this, a comprehensive Government Resolution on JFM was adopted in July 1993 to formalise the arrangement. This granted communities usufruct rights to 100 per cent of NTFP and to 50 per cent of the final harvest of timber. Although this seems like a favourable arrangement, especially with regard to NTFP, in fact villagers were not free to sell their produce to anyone other than NTFP lessees or departmental agencies. Nevertheless, since the 1993 resolution, approximately 1100 *Vana Sanrakhyana Samitis* (VSS: Forest Conservation Societies) have been formed in Orissa, protecting and managing forests covering 0.63 mha.

Objections from those involved in Community Forest Management

While JFM is hailed by many as an important development in community participation in forest management, one interesting and significant aspect has been the reluctance of many communities in Orissa, in particular, to have anything to do with it. This reluctance derives in part from a long history of distrust of the Forest Department by local communities. It also derives from a history of protection and management by independent, grassroots, self-help forest groups like BOJBP, who have devised a process known as Community Forest Management (CFM). The recorded history of CFM dates back to 1936, when the village of Lapanga in Sambalpur District in west Orissa started to protect nearby natural forests. Then in the 1960s and 1970s, many villages in western and central Orissa began protecting their nearby natural forests without the help of the Forest Department. The districts of Nayagarh, Mayurbhanj, Bolangir, Sambalpur, Keonjhar, Dhenkanal, and Phulbani are areas where CFM has made considerable headway. In these districts, in particular, communities are at loggerheads with the Forest Department over the implementation of JFM, believing, as they do, that Forest Department intervention with money and strategies for scientific management will undermine their local initiatives and do more harm than good. They also believe that the JFM partnership between the community and the Forest Department is a very one-sided affair, since the Forest Department holds the real power.

Presently in Orissa, at a rough estimate, there are reckoned to be more than 10,000 CFM village groups, protecting nearly 2 mha of forest. (This conflicts with figures from a survey conducted by the Forest Department in 1995, which reckoned that only 2691 villages were actively involved in it in the State.[18]) Not only are these groups doing local forest protection, but they have also developed their own networks for solidarity and the exchange of information. These operate at village-cluster level, at district level, and in some places at a regional level. As the forest resources grow and mature, conflicts within and between villages are on the increase. Therefore, an important function of these networks has been to resolve conflicts and to fight together for greater rights over forests and their produce.

JFM: new developments
In February 2000, the Central Ministry of Enviornment and Forests intoduced new guidelines to strengthen JFM. These included the following:

- registration of local JFM committees as either societies or co-operatives to give them stronger legal foundation;
- enhanced participation of women in JFM, with half the places on local JFM committees being allocated to women;
- decentralised preparation and approval of micro-planning in JFM areas;
- extension of JFM to 'good forests', whereas earlier it was limited to degraded forests;

- networking among JFM committees, forest officials, and NGOs to improve ways of working and to help, among other things, in the resolution of conflicts;
- legal recognition of self-initiated forest-protection committees (i.e. CFM groups).

As yet, no State governments have modified their JFM orders or passed resolutions in line with these new central guidelines; when they do, however, it is expected that the application of JFM at the local level will be genuinely more community-centred.

Current challenges in forest management

Since the advent of economic globalisation, the major threat to India's forests now comes from the national and international corporate sectors. The threat is gathering pace. Both at the national and State levels, hundreds of new projects for mines, industries, and power generation have been initiated. As a result, the country is threatened with a looming ecological crisis of potentially catastrophic proportions. And first among the casualties are India's forests and forest peoples.

Furthermore these forests are facing growing biotic pressure as they are clear-felled for their timber (most of it illegally), cut over for their fuelwood, encroached by the landless, and over-grazed by domestic animals. Notwithstanding severe reservations about JFM, and despite the government's lack of support for CFM initiatives, both JFM and CFM offer rays of hope. But government, at Union and State levels, will have to reorient its thinking and practices with respect to both systems. JFM needs to be made into a genuine partnership, and CFM needs a policy framework within which communities will be granted more rights over the forests that they protect. Unless this happens, the battle to save India's dwindling forest resources will be lost in the face of unrelenting pressures.

Summary

India has one of the richest biotic endowments of any country in the world. Much of its diversity lies in its forests. But these forests, once very extensive, face unprecedented threats. The origins of these threats lie in the 1860s and 1870s, when the colonial government first claimed the wealth of India's forests as its own, to be exploited for raw materials and for revenue generation, regardless of the fact that millions of people lived in them and depended on them for their livelihoods. A succession of laws was passed over the next century to provide the frameworks for exploitation, production, conservation, and protection. Most of these were strong on exploitation and production, and weak on conservation and protection. Furthermore, they progressively reduced the traditional rights of indigenous peoples, of whom more will be said in the next chapter. It was only in the 1980s that the legislative and policy pendulum started to swing back in favour of forest conservation and the livelihoods of forest dwellers. But while enforcement has slowed the pace of destruction, it has not been strong enough in its support for forest conservation and the protection of forest dwellers. In recent years, following the failure of the Social Forestry Programme, Joint Forest Management has been introduced. Although conceived as a partnership between State Forest Departments and the people, it is regarded with suspicion by many communities and NGOs – in Orissa in particular. For them, the way forward is through Community Forest Management, that is grassroots community action, as exemplified by BOJBP and its sister organisations, to protect what they see as 'their forests'. While CFM has a long history, it is only now beginning to spread widely. As it does so, it pits many village groups against JFM. And because the latest changes in JFM guidelines have yet to be implemented at the local level, this confrontation continues. Meanwhile, with the liberalisation of the Indian economy and its penetration by global market forces, India's natural forests face an assault as never before.

India's forest peoples 3

...the tribals of central India are among the nation's poorest social groups. Due to their isolation and minimal political influence, these communities have had little access to education, health or other economic assistance. Yet, while many tribal people rely on subsistence hunting, forest-gathering, and shifting cultivation for their livelihood, they possess a strong sense of cultural identity and have often been able to maintain an economic independence that has kept them united when facing exploitation by more powerful groups.
(Mark Poffenberger, Betsy McGean, Arvind Khare [1])

Origins and identity

In the mid-1980s the Bhopal-based National Centre for Human Settlements and Environment (1987)[2] estimated that 48 million people in India lived within and in the vicinity of forests, and were dependent on forests for their livelihoods and means of survival. In addition there are another 250 to 350 million who are partially dependent. A very large proportion – at least two-thirds – of those who are most reliant consists of communities known variously as 'tribals', 'Scheduled Tribes', and *adivasis*. These are the descendants of the earliest inhabitants of the country (*adivasi* means 'original people') who were pushed by invaders from the north into the hillier and less accessible parts of the country which the conquerors did not want, or simply did not reach. Tribal peoples are scattered throughout India (see map, Figure 5). Their greatest concentrations are in the forested hills and mountains of the north-east, and in a broad belt of hilly, forested country across north-central India from the eastern side of Gujarat and Rajasthan in the west, through Madhya Pradesh, to the western part of Orissa and south and south-east Bihar in the east. They do not form a homogeneous society; neither are they a single ethnic or linguistic group. In the country as a whole, 212 Scheduled Tribes are enumerated, constituting about 7.5 per cent of the population, that is more than 70 million people.[3] In the past they lived very much in isolation from each other and from mainstream Indian society, most of them choosing forested habitats within which they could live in reasonable peace and quiet. However, their chosen isolation was breached when the British took over India and extended the rule of 'law and order' over them, driving roads and railways through their lands, curbing their traditional rights, and giving access and protection to traders and money-lenders who exploited and ultimately ruined millions of them. What is true for tribal communities is also applicable to non-tribal forest peoples, who over the years have been subject to many of the same destructive forces.

As a result of their exposure and subsequent exploitation, very large numbers of forest peoples now no longer live in forests. Millions have been drawn into mainstream society, becoming labourers, often 'bonded' through debt, on the farms and plantations, and in the mines of those who have displaced and colonised them. With cruel irony, many have also been caught up in the clearance of the very forests that were their own, to make way for mineral and power extraction, commercial forestry, and agriculture. Others have been drawn into alien slums in towns and cities, far removed from their homelands. For them the benign rural jungles that were their birthright, whose ways they understood and with whose natural laws they worked in harmony, have been replaced by hostile urban ones. And all this has happened in spite of, indeed in the face of, Article 36 of the Indian Constitution, which declares:

> The state will promote with special care the educational and economic interests of weaker sections of the people, and in particular the Scheduled Castes and Scheduled Tribes, and shall protect them from social injustice and all forms of exploitation.

A sustainable and holistic way of life

It is all too easy to present a romanticised picture of India's forest peoples, living lives of noble innocence and simplicity. However, in comparison with the more powerful societies of caste India that have overrun them, theirs was – and, where it remains, still is – a lifestyle of extraordinary balance and self-reliance. Broadly there are three types of traditional forest community in India.

Hunter–gatherers

First, there are pure hunter–gatherers, whose numbers now are minuscule, but who in the early years of the twentieth century numbered, according to Madhav Gadgil and Ramachandra Guha,[4] a dozen groups widely scattered from the Himalayan foothills in the north to the forests of Cochin in the south. Such hunter–gatherers as survive today exist largely by making their skills and knowledge of the forests available to the new owners of their

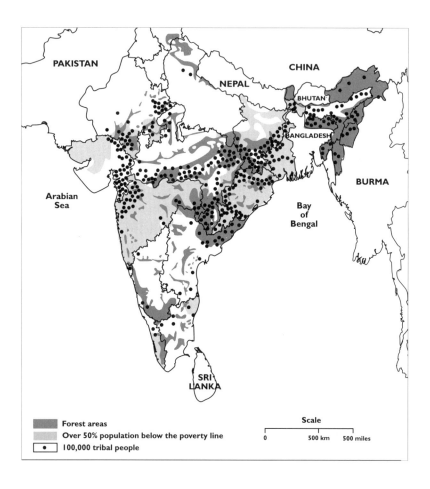

Figure 5: Map of forested areas, tribal areas, and the incidence of greatest poverty in India, showing that there is a broad correlation between the three: forest-dwelling tribal communities are among the poorest in India.
(Source: M. Poffenberger and B. McGean (eds.), 1996, *Village Voices, Forest Choices*, Oxford University Press, Delhi.)

homelands, the local Forest Departments. Some are employed in the capture of elephants, others collect such forest products as honey on behalf of the Forest Departments and merchants.

Shifting cultivators

The second group are the shifting cultivators. They are a large group, of many millions, consisting mostly of tribal peoples, scattered throughout the country, but found in greatest concentrations in the tribal-forest belt, to which we have already referred, across north-central India. Shifting cultivation goes by many names: swidden, slash and burn, *jhum, kumri, podu* (in Orissa), and *bogodo* among many tribal communities. The essential feature of this form of cultivation is, as its name implies, its shifting nature. Shifting cultivators, or *jhumiyas* as they are sometimes called, clear a patch of forest by axe and by fire, cultivate it for a few years (depending on how long its fertility lasts), and then move on to a new patch, returning to recultivate a former patch only after many years of fallow. While shifting cultivation has often been called 'primitive' by mainstream Indian society, in reality it is highly sophisticated, having been developed over several millennia by close study and understanding of the ecology of forests and of the interaction of plants, animals, and human beings within them. Far from being primitive, shifting cultivation is regarded by many academics as a rational form of agriculture in the context of forest environments worldwide. It has many things to commend it. First, it is non-destructive to the forest itself. By operating a long cycle of fallow (18–20 years used not to be uncommon), it prevents over-exploitation of the biomass, and allows forest regeneration. Second, it provides – albeit for only a few years at a time – fertile soils for forest farmers. This fertility is enhanced by ash when the forest undergrowth is initially cleared by burning, and is sustained by the droppings from domestic animals. Third, shifting cultivation protects forest soils: it allows soil fertility to be restored; it limits soil erosion, and it helps to maintain soil moisture and ground water through the protective effects of the surrounding forests. Fourth, by creating minimal environmental impact, it helps to moderate climatic extremes, so reducing the risks of floods and droughts.

One of the most essential aspects of shifting-cultivator communities, and the last remnants of the hunter–gatherer societies too, is that, while they depend on the forest for their livelihoods, they do not exploit it: rather, they are its guardians. In response, nature provides them with an immense variety of goods and materials for their everyday survival: fodder, fuelwood, timber, medicines, bamboo, and a vast menu of foods. An equally important point is that, in order to survive, they need to have a deep understanding of the forest

Community forest management

and its ways. Many may not know how to read and write, but they know an immense amount about their environment. This knowledge has not been learned in schools, but within their communities, handed down, refined, and passed on over the generations. In their book, *Asking the Earth*,[5] Winin Pereira and Jeremy Seabrook tell of a twelve-year-old Warli girl, Raji Vavre, living in the forested foothills of western Maharashtra, who knows the names of more than a hundred herbs and shrubs and their uses:

> She knows which plants are a source of fibre, which are good for fuel and lighting, which have medicinal uses. She knows how to get crabs out of holes and how to trap fish. She can catch wild hare, quail and partridges, and locate birds' nests.
>
> She possesses a vast complete knowledge system, which includes such orthodox divisions as animal husbandry, agriculture, meteorology, herbal medicine, botany, zoology, house construction, ecology, geology, economics, religion and psychology. And, more important, she is also part of a remarkably successful education system: because her father has died, she taught all this to ... [her] younger brothers and sisters ... Moreover, Raji is not unique. Most Warli children of her age have the same or even more knowledge.

A diversity of skills

Like many tribal groups, the Warlis are multi-skilled non-specialists: most know how to cultivate, rear animals, build houses, hunt, fish, gather wild foods and other products from the forest, and process them. Many are also extremely skilled herbalists and artisans.

Let us look at their farming. Most shifting-cultivator societies in India grow a great variety of crops. The Warlis, for example, grow several kinds of pulse and vegetable. They also cultivate many kinds of paddy (rice). Their reasons for this are that different types of paddy have different requirements of water and soil, and have different susceptibilities to pests, diseases, and drought. They also mature at different rates and times. By growing many types, none of which may be as high-yielding as the well-known High-Yielding Varieties (HYVs) of the Green Revolution, they can ensure that whatever nature delivers in a particular season, by way of the weather and infestations of pests and diseases, they are always likely to have something to harvest. For, like all shifting cultivators, they know that specialisation is risky.

In addition to cultivating, many of these communities keep a few animals: cattle, goats, and chickens, which graze in the forests and which are also fed on fodder collected from the forest – grasses, shrubs, and loppings from trees. They treat their animals with home-made herbal medicines, as they treat themselves, when they are sick. These medicines are derived entirely from the forests in which they live.

'Shopping' in the forest

Shifting cultivators also harvest a range of forest foods to supplement the produce from their farming. While these are important all the year round, they are vital in the lean season (March/April to August/September), when there is most likely to be a grain shortage. These wild foods include tubers and corms, leaves, flowers, fruit and nuts, and fungi; and also small game, wild honey, birds' eggs, and fish. Before forest destruction and Forest Department restrictions, wild foods used to be much more important than they are now; for example, at one time the flowers of the *mahua (Madhuca indica)* provided the Warlis with 30 per cent of their total food requirements. After picking, *mahua* flowers are sun-dried, and subsequently used as a vegetable, rich in protein and vitamins. They are also brewed into *mahua daru*, an alcoholic drink, which is rich in both medicinal and nutritive properties. For many forest communities, *mahua* seeds are an important source of cooking oil, too.

For shifting cultivators, the forest is also an important source of building supplies. They use different trees to provide timbers for different parts of the framework. They use reeds and bamboo for the walls, often smeared with a mixture of mud and cow-dung. Floors are made with the same compound, and the roofs are thatched with grasses or straw. When the thatch rots, it is easily replaced and then recycled as compost. Because of forest destruction and restrictions placed on them by Forest Departments, forest communities are finding that there is a growing shortage of the best wood for house frames, and houses are now having to be built with inferior timbers, many of which are susceptible to rotting and termite attack.

The forests supply all the fuelwood needs of the communities living within them, but forest destruction means that the quantity and quality readily available have been much reduced. The 'wrong' fuelwoods smoke excessively, causing sore eyes and throats, and worse: some are carcinogenic. The forests also yield abundant raw materials – bamboo, reeds, leaves, grasses, gums, resins, waxes and dyes – for making such items as ropes, mats, and baskets. People may not use all that they gather, but may barter or sell them for clothes and other essentials which they do not produce themselves.

It is clear from all this is that the forests are central to the lives of shifting cultivators who live within them, supplying them with virtually all their daily needs. A recent author[6] on the subject speaks of forest peoples using their forests much as city dwellers use shops, popping into them when they need something. In support of this notion, he cites a study from West Bengal that

Community forest management

Figure 6: Traditional tribal communities live in a close relationship with their forests, taking from it only what the forests can sustainably yield. In this picture a man from a tribal community climbs a tree to tap it for *dhoop*, or resin, used to make ointments, incense, fumigants, and disinfectant. It is also used in the commercial manufacture of paints and varnish

showed that even in seriously degraded forests, after a few years of regeneration, people were using 72 per cent of the 214 species present. This study also showed that, out of the 109 different items routinely used in village households, only a small number were bought in local markets.

A 'green' spirituality

Over and above the material, functional relationship between forest dwellers and their environment, there is a deeper moral and spiritual interconnectedness. They see themselves as part of the forest, not separate from it. They have a deep respect for its ways, and for all living things within it, and this respect finds expression in their myths and legends, and in their social customs and religious practices. It is through these customs and practices that they ensure that their forest continues to be protected and not over-exploited. Thus, certain trees are sacred (such as *banyan, peepel, sal,* and *mahua*); the harvesting of particular forest products is allowed only during certain seasons; hunting particular animals is banned during gestation; timber and fuelwood harvesting is restricted, for example, to one headload per family per week, or one cart-load per month. And among many forest communities, whole sections of the forest are protected for such purposes as dancing, initiation into adulthood, and burial. Others have 'sacred groves' where the spirits of the forest reside. Powerful myths, particularly creation myths, serve to reinforce these rules and customs.

Forest artisans

The third group which has had a close relationship with India's forests are specialist artisanal communities, such as basket weavers dependent on bamboo; smiths who use charcoal for their furnaces; potters reliant on fuelwood for firing; tanners and dyers dependent on barks and shrubs; and silk makers utilising wild silk cocoons. Forest destruction and restrictions placed on access to raw materials have devastating effects on their livelihoods, as has happened throughout the length and breadth of India. Thus between 1909 and 1938, the Agaria, an iron-smelting tribe in what were then the Central Provinces, suffered a huge reduction in the number of furnaces from 510 to 136, as a result of high taxes on furnaces and diminishing supplies of charcoal. This happened in spite of the fact that the communities they served preferred the softer iron produced by the Agaria to imported (British) iron, which was less malleable.[7]

Settled cultivators

To these three primary traditional groups who live within India's forests, who are mostly tribal, and who have a close relationship with their habitat, we add a fourth: the settled cultivators of caste Hindu India. These, who form a large proportion of India's rural population, number between 250 and 350 million. Many are former shifting cultivators who have settled, and many more are encroachers. They live on the margins of forests and use them for very much the same purposes as those still practising shifting cultivation. However, except for the poorest among them, their relationship is not so close and their dependence not so great. But the forests provide them with many of the same supplementary sources of food, with fuelwood and fodder, small timber, bamboo and other artisanal raw materials, and ingredients for medicines. Although their primary relationship is with the land that they cultivate outside the forests, the destruction of forests in close proximity to their villages or simply the reservation of forests (which excludes them) has had a profound impact on the lives of tens of millions of them, particularly the smaller, marginal farmers and the landless. As we showed in Chapter 1 with respect to Kesharpur, shortages of fodder, fuelwood, timber, water, and a wide variety of non-timber forest produce, coupled with falling water tables, increasing soil erosion, and declining soil fertility, have all had a disastrous impact on rural livelihoods, especially for the poorest sections, making them less sustainable and immensely more vulnerable. Madhav Gadgil and Ramachandra Guha argue that a dimunition of sustained conflict between settled cultivators and their local Forest Departments over the years does not mean that huge numbers of them were any less affected by what was happening to the forests around them than were the shifting cultivators. They cite movements that were expressions of major discontent and anger among such communities, such as Chipko in the Himalayas and Appiko in Karnataka, which are considered in Chapter 4.

The impact of 'progress' on the lives and livelihoods of forest peoples

In the above accounts of forest-dependent communities it has been made clear that their relationships with their forested homelands have become increasingly threatened by forces that lie outside their control, as their forests have been penetrated and sequestrated by outside interests. We now look further at the nature of these forces and their impact on forest peoples.

Deforestation and related processes

First there is deforestation, which, as we have shown, is both systematic (planned) and casual (unplanned and often illegal). Satellite imaging shows that the current forest cover of India is 19 per cent, against the stipulated requirement of 33 per cent necessary for maintaining 'environmental stability'. While the rate of deforestation in India is disputed (in the 1980s it was thought by some to be as high as 1.3 mha/year, which was nearly eight times the 'official' rate), there is no doubt that vast areas of the country have been turned into wastelands through forest clearance and degradation. At present it is estimated that 175 mha of the country are degraded, that is, 53 per cent of the county's land area, and over 11 times the area of Orissa, much of it formerly covered by native forests. Commenting on this critical condition, Oxfam India's 'National Strategic Framework, 1996–2000' notes: 'While the blame for forest degradation has long been laid on the increasing population pressure, it is clear that the orientation and direction of state policy on forest management has itself been the primary reason for the worsening state of both forests and forest dependent people.'

As we observed in Chapter 2, until quite recently India's colonial and post-colonial forest policy developed entirely on the basis of accommodating and serving the interests of large forest industries, hungry for timber and wood pulp, and hungry for land for mono-crop plantations, all at the cost of forest peoples. Thus, the 1952 Forest Policy declared that village communities should under no circumstances be permitted to use forests 'at the cost of national interests' – which were identified with defence, communications, and vital industries. Should anyone doubt what this meant for forest peoples, an extraordinary sentence in paragraph 13 of the Policy makes it clear: *'The accident of a village being situated close to a forest* does not prejudice the right of the country as a whole to receive benefits of a national asset' [our emphasis]. The policy also stressed the need to preserve India's forests, and to adhere to the 33 per cent optimum cover referred to above. As we noted in Chapter 2, it sought to maintain this by discouraging forest-dwelling communities from practising shifting cultivation and uncontrolled grazing.

In that same year – 1952 – India's first Five Year Plan set out the government's objective of converting what were termed as 'low value' mixed indigenous forests, with their immense biodiversity and their social and economic significance for millions of people, into 'high value' plantations of single-crop commercial species, such as teak and eucalyptus. What forestry meant in the context of the Plan, therefore, was the 'production of trees' to meet India's industrial and commercial needs. In order to achieve this objective, there had to be *deforestation* of native forests (and people's homelands), to make way for *afforestation* through the plantation of mainly exotic fast-growing varieties.

From construction to destruction

As, under these policies, vast swathes of India's forests were clear-felled and replaced with exotic plantations, so there was growing pressure on the remaining forests from forest communities to meet their subsistence needs. These included the needs for fuelwood and small timber; for fodder and food from farming forest patches under shifting cultivation; and for a diverse range of non-timber forest produce (NTFP) of the kinds described earlier. That these pressures also grew as population grew is not denied, but the major issue is not one of more people, but of a rapidly shrinking resource base. A study by Fernandes et al., already noted, makes it clear that, as a result of being 'squeezed', forest communities changed from being people who lived *from* the forests in a constructive, protective relationship with them, to people who lived *on* them, in a destructive and unsustainable way.

How was this manifest? It started with a huge reduction in the period of fallow allowed for in shifting cultivation, from 18–20 years to six or even three years, as people were forced to return to their former plots long before traditional practice had allowed. This prevented the forest from ever properly regenerating, which in turn led to a reduction in the quantity and quality of domestic timber, fuelwood, and NTFP. It also led to a huge decline in soil fertility and soil moisture, and an increase in soil erosion. As a result, less food could be cultivated, at the very time that less wild food was available. All of this forced forest communities to depend more and more on cash sales of fuelwood and smuggled timber, for which there were hungry urban markets, usually close by.

At the same time as all this was happening, there was another, external, pressure on the forests. It came from small 'outside' land-holders, whose growing marginalisation and indebtedness to money-lenders forced them to seek new sources of income. For these people, fuelwood and small timber were an obvious resource to exploit; so was forest land for farming by encroachment. Indeed, if they were to avoid migration to the cities, most had little alternative. Meanwhile, the process of indebtedness among forest people themselves led to money-lenders illegally sequestrating the best forest land, causing a further reduction in the resource base of forest communities. In addition to all this, there was the corrupting influence of the contractors (in league with foresters, politicians, and industrialists), whose manifest greed and irresponsibility towards the forests rubbed off on the rural poor, who developed the attitude: 'If that's what they're doing, why shouldn't we?'

Which of the twin pressures on India's native forests, commercial and domestic, has been the greater? It varies enormously from one place to another. One study[8] by a voluntary agency in the early 1980s showed that two-thirds of the rapid deforestation in the northern State of Himachal Pradesh was due to commercial interests, and the rest due to excessive exploitation by the local forest dwellers. In a sense, however, it is pointless to judge which is the greater 'evil', since the one is directly responsible for the other. 'Over-exploitation of forests by people', write Fernandes et al., 'is not independent of industry.' Indeed, the extent to which forests are destroyed by industry has a direct bearing on the extent to which forests are destroyed by people. The one is inseparable from the other.

The loss of non-timber forest produce
In addition to the loss of land, fuelwood, and timber for forest communities through production forestry, and the 'squeezing' that forest communities suffered in consequence, there has also been a direct loss of NTFP through commercial–industrial exploitation. One of the most significant of these is bamboo. Once regarded as a weed, bamboo is now an important raw material in India's huge paper industry. State Forest Departments now offer leases on generously subsidised terms to companies to exploit bamboo forests, regardless of the fact that bamboo is a significant resource for forest dwellers, who use it in making baskets and other household tools, in fashioning musical instruments, and in constructing houses. Although, according to the rules, only inferior-quality bamboo should be harvested for industry, in order to preserve the better growth for artisans, in practice this stipulation is ignored. Also, whereas forest peoples coppice bamboo and other forest plants in a way that encourages growth, industrial methods simply clear-fell, which prevents it. While the assault on India's bamboo forests is nation-wide, it has been particularly savage in Orissa, where some observers believe there has been a 50 per cent reduction in just ten years.

Further damage has been inflicted on those parts of the forests that are rich in *tendu* trees (*Diospyros melanoxylon*). *Tendu* leaves are used in the manufacture of indigenous Indian cigarettes (*bidis*). Frequently in areas where the leaves are collected by non-forest people, undergrowth is cut back, in order to promote better growth. This happens regardless of the fact that within the undergrowth are many food plants and medicinal plants important to forest people.

Two other important NTFP products are *mahua*, whose uses have been described in an earlier section, and *sal* seeds (*Shorea robusta*), which are used by tribal communities to make a highly nutritious buttery cooking medium. There is a rapidly growing demand for both by the urban industrial economy for the manufacture of soap and hydrogenated oil, which means that less is available to those who have traditionally depended on them.

In spite of the fact that the loss of NTFP is a very serious problem for forest communities, the regeneration of NTFP has attracted very little official attention, with little or no place for it in the one place where one would have expected to have found it, that is, in the Social Forestry Programme. In Orissa, where the dependence of tribal communities on NTFP is high, only 4 per cent of the trees planted in 1986–87 were NTFP species, according to the State government. One reason given for this lack of interest is that many NTFP species require a relatively long period at the seedling stage, compared with fast-growing timber species, such as eucalyptus, which can be planted out after only a few months. They also mature much more slowly. Social Forestry staff, who have ambitious targets to meet, are not inclined, therefore, to spend time either on growing them or on promoting them.

Nationalisation of non-timber forest produce

A further problem for forest communities has come from the nationalisation of NTFP, implemented in stages from the 1960s onwards with the intention of raising revenue and of helping the poor, by protecting them from exploitative intermediaries and traders. In reality it has done anything but help them. In their book, *To the Hands of the Poor*, Chambers et al.[9] point out that by reducing the number of legal buyers, nationalisation chokes the free flow of NTFP. It also delays payment to the gatherers, because government bureaucracy grinds slowly and cannot make prompt payments. All this results in contractors entering by the 'back door', taking up unofficial 'franchises' – and these are the very people whom nationalisation aimed to eliminate! Furthermore, because these dealers now require higher margins to cover uncertain and delayed payments, and to ensure that the police and other authorities 'ignore' their illegal activities, they pay the collectors even less than they got before. As far as Orissa is concerned, few NTFPs are nationalised, but most have always been treated as if they were, and the trade has long been in effect monopolistic – that is until very recently.

What happened in 1997 was that the government of Orissa amended legislation in line with that of the central government and vested the ownership of a range of 'minor forest produce' in *gram panchayats* (local councils, usually comprising clusters of villages). But the implementation was delayed, and it was not until March 2000 that the law became reality, and the government finally defined control and ownership. It divided NTFP into four categories, and declared that control of one – minor forest produce (MFP) – would lie in the hands of the *gram panchayats*, who would be free to decide the mechanism for trading it. According to this four-fold categorisation, MFP does not include certain nationalised products such as *tendu* leaves and *sal* seeds, and certain other products (such as bark) the harvesting of which risks damaging trees. But it still covers 60 items that can be gathered and traded without any royalty being paid to the government, and without any transit restrictions being imposed. *Panchayats* will have no rights over MFP in Reserved Forests, unlike tribal communities and certain artisans.

The most important aspect of this policy is that it is applicable to the whole of Orissa, so that all *gram panchayats* in the State now have control over 60 items of minor forest produce, which are now free of monopoly trade rules. Because this announcement is very recent, it is too soon to judge its impact; but on paper it represents an enormous step forward for the rights of forest-dependent peoples over forest products.

The impact of the Social Forestry Programme on the rural poor

In Chapter 2 we looked at the rise and fall of Social Forestry. We noted how the whole programme favoured the farm-forestry component of the programme, and how it was subverted in favour of the larger farmers and forest-dependent industries. We saw that although it had the effect of reducing the net rate of deforestation, it did not significantly reduce the pressure on the natural forests, and it did not benefit those who most needed it: the rural poor. It did not satisfy their needs for fuelwood, fodder, and small timber, but it did enhance the income of the larger farmers. Such afforestation of degraded community lands as did take place was undertaken largely by the local Forest Departments, and not by local people. And although Village Forest Committees were set up within the framework of existing Village Councils, they were dominated by the local Section Forester, who became the member-secretary. Villagers on the committee essentially became passive observers, and the rest of the village felt entirely excluded. There was, therefore, no sense of local ownership or stewardship of these lands.

There were other adverse effects as well. Because so much farmland was made over to tree growing, there was a reduction in food crops, and therefore less food was available in local markets, which severely affected the poorest families. There was also an increase in rural unemployment, since trees need less labour than crops. A further negative impact was a reduction in fodder, since farm trees are usually grown so close together that grass does not grow under them. Also the loppings of eucalyptus are unpalatable to domestic animals. As we observed in the last chapter, in the terms in which it was conceived, Social Forestry was 'a disaster in every possible way'.

Big schemes

The national context

After deforestation, the second major threat to the lives of forest-dependent communities comes from large-scale development projects, and the construction of the infrastructures that go with them: roads, railways, and townships, which demand not only land but timber too, such as sleepers for railways and timbers for sheds. It is reckoned that more than 45 million people have been displaced in rural and urban India since Independence by large development schemes of one kind and another: that is nearly one million people a year. It is further reckoned that, since the rate is rapidly rising, within a few years the numbers annually displaced will reach two million. The proportion of people displaced from forests is not known but is considerable, for the simple reason that many of these schemes are located in forested areas. The reason for this is that beneath the forests of India, particularly those of Madhya Pradesh, Bihar and Orissa, lies immense mineral wealth (bauxite, chromite, iron ore, manganese) and huge reserves of coal. In all of these States, as mines have been opened up, millions of tribal and other forest peoples have been displaced, the vast majority without any compensation. Also through these forests flow some of the great rivers of India, many with enormous hydro-electric and irrigation potential. One of the greatest of these is the Narmada River, whose story is told in the next chapter.

Orissa

What is happening in Orissa with respect to displacement by big projects is a reflection of what is occurring countrywide. In the south-western district of Koraput, for example, 18 large projects – consisting of mines, dams, and wildlife sanctuaries – have displaced no fewer than 100,000 people, mainly Scheduled Caste and tribal communities. The giant Hirakud Dam in the north-west has displaced 285 villages and more than 22,000 families. Some of these Hirakud oustees have been displaced three times: by coal mining, by a thermal power plant, and by the dam itself. The Balimela Dam has displaced a further 2000 families, the Upper Kolab more than 3000, and the Rengali in excess of 10,700. While some of these development projects have incorporated resettlement schemes, most have not. Even where they have done so, the compensation packages have been utterly inadequate to compensate for loss of land, and the destruction of livelihoods, way of life, and culture. In Chapter 4 we look at the response of local communities to some of these projects.

In terms of compensation and resettlement, shifting cultivators are particularly vulnerable. Since they have no individual rights to land, there is no basis, as far as the government is concerned, on which to assess compensation packages (land for land or cash for land). Even where land for land is granted, it is forgotten that forest communities often derive more than 50 per cent of their sustenance from the surrounding forests in the form of NTFP. 'Rehabilitation authorities only think of the amount of land being cultivated by persons displaced by a project', write Fernandes et al.[10] 'The whole infrastructure that sustains them is forgotten.'

Whatever the outcome for forest people of big developments in their forests, whether they are compensated or resettled or not, they are displaced, and that displacement is not only destructive to them as a people: it is usually destructive to yet more forest.

The impact of economic liberalisation

Liberalisation of the Indian economy and the fast-track race for development being pursued by many States has significantly increased the rate of displacement, since it has greatly accelerated the granting of licences for mining and other industrial activities. But at the same time as the drive to initiate more big projects, there is increased mechanisation in existing industries to achieve higher productivities.[11] This is leading to a huge 'shedding of labour', as it is euphemistically called, much of it in the form of people who have already suffered displacement. And, of course, many of these are from Scheduled Tribe and Scheduled Caste communities, who thus pay a double penalty.

Violation of the rights of forest communities

A third threat to the lives of forest communities comes from the definition and redefinition of forests (and people's access to them), and the failure of the authorities to help forest communities to understand their rights and privileges. Even though certain rights of forest communities were established by early colonial policies, and have not been directly countermanded since those days, nevertheless *de facto* these rights have been curbed or deliberately shrouded in mystery. First, State governments have re-designated many areas of forest and changed local people's rights to them. Significant areas of forest that were 'protected' have become 'reserved', and when this has happened people's rights have been much reduced. Second, State governments have created new rights for industrialists (for example, to harvest bamboo for the paper industry), which frequently supersede the rights of forest peoples. Third, forest peoples are, in the understated words of Chambers et al.,[12] 'far from fully informed about what they can legally collect from forests and what is prohibited'. Various surveys have been conducted to find out what people know of their rights. They reveal confusion at best, and at worst complete ignorance. It almost goes without saying that it suits both Forest Department officials and traders to keep it that way: the more ignorant or confused that people are, the less likely they are to assert their rights and the less likely they are to fight the exploitation that is perpetually visited upon them.

The whole issue of forest rights is, of course, dependent upon the modern concept of ownership, which is completely opposed to that of the forest dwellers. As we have argued in this chapter, for tribal communities the forest is central to their lives at every level: material, emotional, social, and spiritual; but for the State it is simply a *place of extraction* to feed the ravenous needs of industry for raw materials, and the needs of the State for revenue.[13] It was for this reason that in the 1860s and 1870s forests were 'reserved' and put under the control of the State Forest Departments. Once this happened, forest dwellers were viewed as 'enemies' of the forest whose traditional rights and practices ran counter to those of the State. Within this new legal and policy framework, any rights which forest communities were allowed were to be seen not as rights *per se*, but as concessions or privileges, which could be withdrawn at any time.

The impact of deforestation on women

Deforestation affects women much more than men, and the poorer they are the worse it is for them. Although in traditional forest communities, particularly tribal ones, there has often been a greater degree of equality between men and women than in mainstream Indian society, there still has been an unequal division of labour. Thus, in the late 1980s, Fernandes et al.[14] found that tribal women in Orissa played a major role in the economy, working on average three hours a day longer than men, a workload that significantly increased with deforestation. Tribal women have been traditionally involved in collecting water, fodder, fuel and other NTFP, while the men have done most of the cultivating and hunting. With deforestation, women's work of fetching and carrying becomes more difficult, since they have to go farther and farther from their villages to reach the receding tree line. One study in Orissa[15] found that, over a twenty-year period from the mid-1960s to the mid-1980s, the average distance that people (mainly women) had to walk to collect firewood, bamboo, fodder, and other products increased from 1.7 to 7.0 km. Other studies have shown that the situation deteriorates over time: while working longer (often up to 14 hours a day), they collect less, and their lives become even more difficult. A study,[16] quoted in Chambers et al., in one area of South Bihar in the early 1980s, described how every day

300 women went into the forests to collect firewood from illegally cut timber. They earned Rs.120/month, an amount so paltry that half of them were permanently in debt. To reach the forest, they walked as far as 12 km and then, when they had finished collecting wood, travelled by train with their head-loads to town. During the whole process they were obliged to bribe the village headman to allow them to do it, the forest guard to look the other way, and railway staff to allow them to travel 'free' on the train. Hardly surprisingly, they were not left with much profit at the end of it.

One obvious outcome of all this is that women have less time to take care of themselves, even when they are ill. Several studies in India and elsewhere have shown that, in situations of social and economic disintegration, the proportion of men to women attending primary health centres is five to one. This is in spite of the fact that women are likely to be less healthy than men, because they are less well-nourished and are working so hard. They are also less likely to have to hand the pharmacopoeia of herbal medicines that was available before forest destruction.

The impact of male migration and the commercialisation of NTFP

Women's burdens are further increased when the economic situation within their families becomes so severe that the men are forced to migrate in search of work, be it seasonal or permanent. When this happens, the women end up trying to do the cultivation in addition to all the fetching and carrying, the housework, the mothering, and the caring for the elderly. In many cases this proves to be an impossible burden. In order to survive, women may have to borrow from a money-lender, to whom they become indebted, and to whom they may eventually lose their land, even though it is illegal for non-tribal people to own tribal land. Even if it does not go this far, uncultivated fields are at risk from encroachment by others in the community. This is something that would never have happened in the past, when the community's values were still intact and people helped each other in difficult times.

Even apart from such extremes, the position of women can be severely undermined simply through the commercialisation of NTFP, which makes gender relations more inequitable. Mitra and Patnaik, authors of the recent Oxfam report cited earlier,[17] point out that before commercialisation women decide many things about the collection and allocation of NTFP, but commercialisation takes away their decision-making power and puts it in the hands of men – a situation which can spell disaster for the women.

It is not difficult to appreciate that, while everyone in forest communities is affected by deforestation, it is women who suffer the most.

Deteriorating health and nutrition

Nor is it difficult to understand how deforestation has led to a deterioration in the overall health of forest communities. Declining yields of food crops from the forest, and increasing indebtedness (which forces people to sell more of what they gather and grow) have reduced both the quantity and quality of food available to forest people. Among the tribal peoples of Orissa, one piece of research[18] showed increases in night blindness, dental caries, anaemia, anorexia, bleeding gums, and many other conditions. Also recorded are increases in stomach disorders caused by contaminated water sources, and increases in malaria among people living near dams and canals. In addition to all this, there has been a significant loss of sources of traditional herbal medicines available from the forest, a loss that has not been made good by increased access to State health-care provision, which for many communities is still non-existent. The result is that there are high levels of morbidity and mortality among many communities affected by deforestation. It is reported that deaths due to the diseases of old age are not common, since diseases like malaria, tuberculosis, and asthma take their toll at earlier ages. Even deaths due to starvation have been reported.

Summary

The numbers of people who depend to a greater or lesser degree on India's forests are in the hundreds of millions. Of these, around 50 million have a very close relationship with those forests, using them for farming (traditionally shifting cultivation); for timber, fuelwood, and fodder; for a vast range of edible produce (fruits, nuts, tubers, honey, and game); for ingredients for herbal medicines; and for raw materials for house construction, craftwork, and tool making. A significant proportion of those who are most dependent are the tribal peoples – the aboriginals of India. Their relationship with the forests is more than just a functional one, having an important moral and spiritual dimension. They see themselves as being both part of the forest and the guardians of it, and they have important rules of forest protection, powerful myths, and deeply held spiritual beliefs.

However, since India's forests have been seen by successive governments as resources to be exploited to help to create a modern India, rather than as habitats to be conserved for the sake of those who live within them, then it follows that the needs of India's forest peoples have long been subservient to the primary purpose of national development. Through successive plans, policies, and legislation, Indian governments have consistently spelled out their aims with respect to forests as being on the one hand to serve the nation's defence, communications, and industrial needs, and on the other to restrict the rights and usage of its forests by the millions of peoples who actually inhabit them. Under the banner of scientific or production forestry, this has resulted in massive deforestation (with illegal felling now being much greater than legal clearance); the creation of extensive mono-crop plantations; and the loss of rights, land, and resources of forest-dwelling communities and communities living close to forests. Millions have also been affected by the exploitation of other resources lying within India's forested areas: minerals, fossil fuels, water, and water power. Big schemes now displace well over one million people a year, the vast majority being people who live in forested domains. It is predicted that within a few years this figure will reach two million per annum.

The consequences for forest peoples have been catastrophic: the destruction of their habitats; the loss of their livelihoods; the break-up of their communities; the virtual enslavement of individuals and families through debt bondage; deterioration in the status and security of women; and a decline in standards of nutrition and health. In addition to all this, there are the environmental consequences of rapid deforestation: soil erosion; loss of soil fertility; falling water tables; and ultimately, unless remedial action is taken, the creation of vast areas of wasteland.

The people fight back 4

'We got only a little food from our fields; when we could not get wood to cook even this paltry amount, we had to resort to a movement.'
(A peasant farmer in Badyargargh, in the hills of Uttar Pradesh[1])

Introduction

The environmental consequences of modern development practices were first brought formally to public notice during the Stockholm Conference on Human Environment in 1972. This conference served to highlight the importance of natural resources for human survival and also helped to inform and articulate a 'pro-poor' environmentalism. From around this time, environmental activists began to politicise communities adversely affected by development projects (dams, mines, industries, lumbering) and started to lobby governments and funding agencies on issues of environmental damage and inadequate compensation and rehabilitation packages. In the 1980s the struggle for environmental protection became more broad-based, with support from national and international environmental groups. This resulted in a burgeoning of protest movements which have challenged the big players that exploit precious natural resources at the cost of the rural poor. Others who became involved were non-government organisations (NGOs), voluntary organisations, local environmental groups, and academic and research institutes, all of which started to advocate changes in polices to support environmental protection and human rights, and to safeguard environmental standards in all development projects. Now there is also perceptible evidence of environmental awareness and concern even in the judiciaries of many countries, where writs, petitions, and public-interest litigation have challenged government and corporate actions.

Popular struggles for environmental protection now encompass movements fighting for the protection of the natural environment, for traditional rights over natural and livelihood resources, and against development projects which displace human habitations. In the following sections, various people's environmental movements in India are considered, beginning with two movements of national and international importance.

One is *Chipko*, the first organised people's environmental movement in India, which had significant success in protecting the sensitive forest eco-system in the Himalayas. The second is the on-going struggle against the construction of a series of dams on the Narmada River, a campaign that has gained enormous international support. Later in this chapter we consider pro-poor environmentalism in Orissa and present a number of case studies.

The Chipko movement: braving the logger's axe[2]

One of the best-known environmental movements in the world, *Chipko* began in the early 1970s as an attempt to protect the precious forests of the great Himalayan region, with its high mountains and deep valleys, exquisite flowers, lush mountain pastures, and inaccessible glaciers. Since 'scientific' forest management by the government had manifestly failed to preserve the forest resources of the area, catering as it did to commercial extraction, the people of the region, who are particularly poor and dependent on the forests, raised their voices against the destruction that was going on all around them.

The movement started in an obscure co-operative organisation formed by a group of young activists in the northern hills of Uttar Pradesh (UP). The organisation was called *Dashauli Gram Swaraj Sangh* (DGSS: Dashauli Village Self-Rule Group). When in 1973 the State Forest Department allotted a huge area of hornbeam forest to Symonds & Co., an Allahabad-based sports-goods manufacturing company, in the heart of the Alakananda valley, DGSS decided to make a stand. It began by organising protest demonstrations in several places in the area. But when these failed to deter the loggers, Chandi Prasad Bhatt, the leader of the organisation, hit upon the brilliantly innovative idea of preventing the felling of trees *physically*, by hugging them. *Chipko* means 'hug' or 'hugging'.

The idea found ready support among the people of the area: even though most villagers in these hills are illiterate, they had started to understand the hitherto unsuspected links between deforestation, landslides, and floods. They now saw, for example, that the flash floods of 1970, which did enormous damage to the sensitive ecology of the Alakananda valley, were in no small measure an effect of large-scale deforestation in the 1960s.

The incident at Reni

A turning point for the movement came when the government auctioned a large area of the forests in the region, and some 2000 trees were to be felled in the Reni forests alone. The contractor arrived to cut the trees while most of the men of the village and the leaders of DGSS were away. But under the spirited leadership of Gouri Devi, the head of the village *mahila mandal* (women's group), the women of Reni hugged the threatened trees and prevented the felling of a single one, and by so doing wrote an important chapter in the history of women's participation in environmental protection.

The Reni incident attracted the attention of the Uttar Pradesh State Government, which instituted a committee to investigate the matter. The committee concluded that it was indeed because of widespread deforestation in the catchment of the Alakananda River that the disastrous floods of 1970 had occurred. A complete ban was placed on all commercial felling in the upper catchment of the Alakananda River and its tributaries for ten years. Another committee was set up to investigate the tapping of resin in the area, and concluded that the rules were being brazenly flouted: trees were overtapped and died, adding to forest destruction. The leasing of forests to the paper mills was also reviewed. Finally a forest corporation was constituted to take over all forms of forestry exploitation in the area.

After Reni

That was not the end of *Chipko*. Over the next ten years, the movement spread rapidly from Garwhal to other parts of the Uttar Pradesh Himalayas. In more than a dozen separate incidents, villagers successfully stopped local felling, and disrupted the auctioning of forest leases to logging companies.

While *Chipko* remained very much a people's movement, its successes were due also to the quality of its leadership, just as with BOJBP. The two leaders of the movement, Chandi Prasad Bhatt and Sunderlal Bahuguna, became internationally renowned, Bahaguna in particular because of his epic fasts, through which many battles were won. For example, to prevent tree felling in the Amlgaddi forest, on 9 January 1979 Bahuguna launched a fast-unto-death, while more than 3000 people, including a large number of women and children, braved the might of the forest contractor and forest officials. Bahuguna was arrested and put in Tehri jail on 22 January. But that failed to deter the determined protesters, and the government had no option but to suspend all felling operations in the area.

Following the successes of *Chipko*, the *Uttarakhand*[3] *Sangharsh Vahini* (USV: The Force for the Struggle of Uttarakhand) was formed in the Kumaon area; it built up a powerful protest movement against all deforestation throughout the UP hills, through mass mobilisation and demonstrations. As a result of their activities, tree felling in the entire catchment area of the upper Ganges was massively reduced.

Believing that this was not enough, however, in April 1981 Bahuguna took the decision to fast again, this time for a total ban on tree felling in the

Himalayas above an altitude of 1000 metres. In response, the government instituted a committee to prepare a comprehensive forest policy for the Himalayas. After the production of the report, the government imposed a moratorium on all commercial felling throughout the whole Himalaya hill zone for 15 years. This brought commercial forestry in the area to a standstill and earned world-wide recognition for the *Chipko* movement.

Strengths of the Chipko movement
Impressive though its achievements have been, *Chipko* should not be viewed as an isolated phenomenon. In many ways it was a mere continuation of earlier social protests in the area. History records various protest movements launched by peasant peoples demanding their traditional rights over forest produce. For example, in the 1960s, *Sarvodaya*[4] leaders organised processions and pickets against widespread distillation and sale of liquor in Uttarakhand. There were also several earlier spontaneous movements in protest against commercial forestry. But because it was the most recent, and because its actions were so inspirational, *Chipko* has become the most famous.

There is no doubt that the leadership provided by Chandi Prasad Bhatt and Sunderlal Bahuguna has been very important for the success of the movement. Bhatt is famous for his reconstruction work and appropriate technology to develop forest resources. Bahuguna is a pure (one might even say 'purist') conservationist, fighting against modern development, working in a prophetic mode. He spreads the message of environment conservation through his strenuous *padayatras* (protest marches), fasts, and mass meetings. He is the person who has made *Chipko* an international phenomenon, and he is the person who has inspired and provided powerful moral authority to many other environmental movements in India. Wherever there is a people's struggle for environmental protection, he has gone there and encouraged them through the message of *Chipko*. (As we shall see in Chapter 5, his visit to Kesharpur in 1983 had a profound impact on many in BOJBP.)

In addition to the personal leadership of Bhatt and Bahuguna, USV's leaders in Kumaon have also been of great importance. Their orientation is different from that of Bhatt and Bahuguna, having been influenced by socialist thinking. Their approach has been to fight for socio-economic redistribution, which, according to them and to other like-minded environmental groups in India, is a prerequisite for ecological harmony.

Chipko has directly inspired a movement opposed to Forest Department practices in the District of Uttara Kannada in Karnataka. Known as *Appiko*,[5] it has successfully campaigned to prevent deforestation by hugging trees threatened by the loggers' axes. It has also persuaded loggers' labourers to take an oath in the name of a local deity to forego tree cutting.

Free Narmada Movement[6]

The largest westward-flowing river in India – 1312 km long – the Narmada is the lifeline of many thousands of communities who live along its course. Because the valley is one of the heartlands of Indian civilisation, there are numerous temples and religious places, which give the valley its unique cultural and spiritual status. From time immemorial, Indians have undertaken pilgrimages to these sacred spots. Living within the valley, specifically within the area of the Narmada Valley Development Project, are seven distinct tribal groups, whose civilisation goes back at least to 1600 BC. With its fertile black-cotton soils, its perennial water sources, and its abundant ancient forests, the valley is very rich in bio-diverse resources for sustaining life.

Damming the Narmada River for irrigation was discussed long before Independence, but it was not until 1961 that Prime Minister Nehru laid the foundation stone, and not until 1987 that the construction of the immense Sardar Sarovar Dam, the keystone of the project, began. When complete, the Narmada Valley Development Project will have constructed 30 large dams and well over 3000 medium and small dams. It is estimated that the total project, if implemented in its entirety, will displace around two million people. While most of the Narmada Basin is situated in Madhya Pradesh, the giant Sardar Sarovar Dam is sited just inside Gujarat. It is a wall of concrete, 1210 metres long, designed to impound an immense reservoir which, when full, will drown 910 sq km of land. The main canal leading from the reservoir will be 460 km long, eventually reaching Rajasthan. A network of secondary canals, totalling 75,000 km in length, will deliver irrigation water to millions of farmers. In 1985 the World Bank agreed to lend $450 million towards the costs of the dam's construction.

Opposition to the dam

Sporadic struggles against the project started in 1979, but it was not until 1986 that opposition began in earnest, led by Medha Patkar, a 'Joan of Arc' figure. Arriving in the valley for the first time in 1985 as a social activist and researcher, she joined the growing protest movement and started to fight for fair compensation for the 'oustees'. It was because of her oratorical and organisational skills that many local people and external activists joined the anti-dam movement.

In 1986 in each of the three States affected by the project (Gujarat, Madhya Pradesh, and Maharashtra), organisations were formed to fight the cause of the oustees, to analyse the government's claims of the scheme's benefits, and to highlight its environmental and human costs. In 1989 the fast-growing movement was united into an alliance, the *Narmada Bachao Andolan* (NBA: Free Narmada Movement).

In 1987 and 1989 Medha Patkar went to Washington to lobby the World Bank to withdraw from the project. She gave testimony to a Congressional Sub-committee, which led the Bank to institute an independent review of the project. A subsequent visit by Medha Patkar to Japan led to the withdrawal of the Japanese government's contribution to the dam's turbines.

On 25 December 1990, 3000 oustees and their supporters set off from the town of Rajghat to the dam site, an event that attracted widespread media coverage and made Medha Patkar into a national celebrity.

In June 1992 the World Bank published the findings of its review, which vindicated many of the claims of the movement. Nine months later, as a result of huge international pressure, it withdrew from the project. This did not deter the Indian government, however, nor did it stop the police from harassing the protesters. In May 1994 the NBA filed a case against the project in the Supreme Court. The Court ordered the government to make public its own internal review and ordered an investigation into the viability of the project.

Although the movement has not succeeded in its primary objective of halting the construction of the dam, it has achieved many lesser but significant victories and in the process has become the most vociferous anti-dam movement in the world. This reputation has been achieved because of its strong and articulate leadership; its powerful and logical reasoning based on careful analysis; its innovative ways of protesting; its adherence to non-violence; its strong media support; and its skilful national and international networking. It has also inspired and given direction to many other environmental protests in the world where people are pitted against modern schemes which threaten the lives and livelihoods of local communities.

Orissa: the fight against dams and displacement[7]

Orissa has been an important battlefield for many environmental conflicts. Whether it is displacement by dams, degradation of the coastal ecology, pollution from industries and mines, deforestation and the destruction of wildlife, or the commercialisation of agriculture and forestry, the people of the State have an impressive record of indigenous environmental movements. Some of the major ones are chronicled below.

Dams have seriously affected the environment and communities in many parts of Orissa. So far seven large dams, 20 intermediate-sized dams, and one small dam have been completed in the State, and 12 large and 11 intermediate dams are under construction. In the absence of holistic policies for resettlement and rehabilitation, and above all in the absence of political will, in most cases it has been a story of blatant disregard for project-affected peoples. In the cases of some bilateral projects (that is, projects supported by a single foreign funder), good rehabilitation packages have been agreed, but they have not been implemented properly. As a result, displaced communities languish in a state of limbo, created by official apathy, bureaucratic delays, and political betrayal. But the experience of earlier projects has sensitised people to raise their voices against any development project that displaces human habitation. Two cases of people's struggles against dams in Orissa, out of several that could be told, are now described.

Anti-Hirakud Dam movement: setting the trend

Construction of the first major hydro-power project in Orissa at Hirakud was initiated in 1946 on the upper Mahanadi, in Sambalpur District close by the border with Madhya Pradesh. The project required 75,984 ha of land, and threatened displacement to 17,393 families in 249 villages. The compensation package was derisory: the total land-for-land award was a mere 10,117 ha, and cash-for-land was a paltry Rs.84 to Rs.100 per acre (0.4 ha). After more than 50 years, many of the displaced people have still not received compensation. At the outset some 4000 or so families accepted what was on offer, mean though it was, but the rest continued to struggle for as long as they could against the construction. But in 1955, when the reservoir started to fill, they were all forced to leave their villages, with or without compensation. Initially they all chose to re-settle near the edge of the reservoir, and for that they received temporary land *pattas* (legal documents). However, these were never finally settled in their names, and still have not been.

The story does not end there, however, for in the area where many of the oustees had settled there are abundant coal reserves, which subsequently became the focus of the government's attention for thermal power generation. So, already once displaced, some people were now threatened with displacement for a second time. But while two villages were vacated, others put up a spirited resistance, as a result of which, and bowing to public pressure, the government agreed to provide homestead land measuring 1 acre (0.4 ha) to each oustee family. This may seem a meagre amount, but it was an improvement on what was initially offered, and it set a precedent for later environmental struggles in Orissa.

Even that is not the end of it, however, since some families were subsequently moved a third time, when the thermal power station was expanded. Yet others face eviction as a result of further expansion now.

The anti-Upper Indravati Dam movement

This multiple-dam project, supported by the World Bank, displaced 4201 families belonging to 96 villages in the south-western Districts of Koraput and Kalahandi when a large area of land was submerged. The first phase of displacement started in 1989, with a meagre compensation package: while the market rate for one acre of land was Rs.30,000, people were paid only Rs.300–600. And the monthly payment of Rs.500 that was promised to each family towards maintenance was never made.

In 1990, the *Indravati Gana Sangharsha Parishad* (IGSP: Indravati Mass Movement Council) was formed and supported by many local groups. Regular meetings, workshops, seminars, rallies, and protests were organised. In June 1990, a unique 'rally' of 5000 domestic animals took place. Women participated actively in the protests. On 5 March 1991 a group of 20 of them decided to undertake a *Jalsamadhi* (mass suicide by drowning). Concerned that this would cause widespread civil unrest, the government was forced to accept at least some of the demands of the affected communities.

In 1992 massive demonstrations took place, as a result of which work on the project came to a standstill. Even the workers on the project stopped work and participated. It was at this point that the government resorted to repression. The leaders were arrested and the demonstrators were harassed. However, this did not deter them. The NGOs and IGSP contacted the World Bank, to protest against the inadequate rehabilitation package on offer. Coincidentally, an accident at the project site led to the death of 16 workers. The World Bank, dissatisfied by what it saw as the negligence of the authorities, stopped its funding, though later did agree to resume it following specific, time-bound commitments made by the government for the proper settlement and rehabilitation of the oustees.

Orissa: coastal environment vs. development

The coast of Orissa consists of sandy beaches, saline estuaries and creeks, lagoons, and brackish swamps with tidal forests and mangroves. Inland lies the coastal belt, the rice-bowl of the State and also its most populous area. It has a special place in the State's economy because of its importance for agriculture, aquaculture, shipping, and tourism. However, both the coast and the coastal belt face increased human interference, which, together with siltation and high tides, has made the whole region vulnerable to severe environmental damage. It is also prone to cyclones.[8] There is a long history of popular struggles to protect the area. In the 1880s people revolted against the colonial government's decision to introduce a canal system for regulating the flow of river water, which local people knew would not work. They contended that their main occupation – agriculture – would be badly affected. They were subsequently proved right, for the area became subject to even worse floods and severe water-logging. A series of protests led to the appointment of a commission of inquiry by the colonial government. But nothing changed.

In recent years people have fought against the government and industrial interests to save the coastal environment. We consider two of these struggles: firstly a protest against the National Test Range, Baliapal, and secondly a campaign against commercial aquaculture in Chilika Lake.

Baliapal: the National Test Range (NTR)[9]

Baliapal, located in the north-east corner of Orissa in Balasore District, is among the most densely populated rural areas of the State. In 1985, the Chief Minister of Orissa announced that a National Test Range (NTR) would be set up there to test and launch missiles. The announcement caused consternation in the area, and at different levels people started mobilising public opinion against the Range. So effective was this that, by the time that the final decision for the location of the Range was made in May 1986, people had already organised themselves into various groups to fight it.

Eviction notices, issued by the government to people in 54 villages in Baliapal and Bhograi blocks, were met with fury, and had the effect of bringing the various groups together into a single Anti-missile Base Committee, spear-heading an Anti-missile Base Movement. The committee immediately set about organising disruption in the construction of approach roads and telephone lines, a blockade of government offices, and the non-payment of taxes.

The area selected for the NTR is highly fertile and enormously productive. While the NTR itself would need only 100 sq km of land, it also required a buffer zone of 300 sq km surrounding it to be cleared for security reasons. This would displace a total of 12,000 families from 126 villages. There was also concern that the pressure waves from the missile firings would be a serious threat to the well-being of people and livestock in the area and that there would be serious atmospheric and water pollution.

The protesters made it clear that they were opposing not the setting up of the NTR *per se*, but the decision to locate it in a fertile and densely populated area like Baliapal. They also feared that the government would not compensate the oustees adequately – an entirely reasonable fear, in view of its poor record of compensation elsewhere. In 1986 a leading human-rights group in Orissa, *Ganatantrik Adhikar Surakhya Sangathan* (GASS: Organisation for the Protection of Democratic Rights), sent a fact-finding team to Baliapal, whose report was widely circulated among government officials, environmental activists, development organisations, and institutions working for human and democratic rights. Following this, many key political figures, including two former Chief Ministers of Orissa and one ex-Union Minister, expressed total solidarity with the Anti-Missile Base Committee.

The situation to date is that because of the strong resistance by the local people and the solid support of democratic and human-rights groups, the National Test Range has not been installed. However, the government has not completely withdrawn from this massive Rs.40 billion project.

The Save Chilika Movement[10]

Chilika, the largest brackish-water lake in Asia, is situated on the south coast of Orissa, some 60 km south-west of Bhubaneswar. The area of the lake, 1055 sq km in summer, swells to 1165 sq km in the monsoon. It is rich in biological species, and part of it has been declared a wildlife sanctuary. It is reckoned that more than 190,000 people depend on it for their livelihoods and survival.

The lake is being severely degraded: its area, depth, and salinity are all declining, as a result of which between 1985 and 1992 fish production decreased by nearly 50 per cent. Although the government is a party to the Ramsar Convention to protect wetlands of international importance, no serious step has been taken to protect Chilika. In fact the very opposite has happened. In 1986 the State government decided to lease 1400 acres of the lake for aquaculture to the Tatas (one of the largest business conglomerates in the country) and took all possible steps to curtail the traditional fishing rights of local fishing co-operatives.

The plan was for Chilika Aquatic Farms Limited, a joint venture of the government of Orissa and the Tatas, to spend Rs.200 million on prawn culture in the lake. The project envisaged the creation of a huge artificial lake inside Chilika by embanking 1400 acres with a 13-km ring dam. The plan was to divide the lake into small ponds for intensive prawn culture and pump into them water from the lake. Using the right protein food, fertilisers, and pesticides, the prawns would be grown to 250–300 grams within 40 days. When the cycle was completed, the polluted water of the ponds would be pumped back into Chilika, regardless of the damage that would be caused to its fragile eco-system.

When the Tatas began construction, a group called 'Meet the Students', from Utkal University in Bhubaneswar, sent a team to study the issue, after which the Chilika Protection Council was formed to mobilise public opinion throughout the State against the scheme. In September 1991, more than 8000 fishermen demonstrated before the State Assembly in Bhubaneswar and submitted a memorandum to the Fisheries Minister. In response, a high-powered committee was set up, led by the Chief Minister.

However, as so often happens in situations like this, the government was slow to act. The protests continued and eventually crystallised into the *Chilika Bachao Andolan* (Save Chilika Movement). During February and March 1992, agitators occupied the land acquired by the Tatas and demolished the embankments. With the support of the police, the Tatas made a bid to repair and reconstruct the dam. The police unleashed a virtual reign of terror to break the back of the protest movement. But, undeterred, the people continued their demonstrations and in the end were able to persuade the government to halt the project. However, unwilling to withdraw completely, the Tatas came forward with a new design, which they claimed would create less environmental pollution. But the people were not impressed and continued to protest.

A major turning point came in 1993, when, in response to a public-interest suit filed by 36 fishing co-operatives, the Orissa High Court instituted a committee to investigate the matter. After the inquiry the Court ordered the demolition of all bunds created in and around Chilika for aquaculture, both those of the Tatas and those of local entrepreneurs. In spite of severe opposition by some business people, the constructions of the Tatas were demolished. Successful cases have now been filed in the Supreme Court of India to ban prawn farming completely in Chilika and along this coast, but the State Government has pleaded for reconsideration of the case. In the meantime, some prawn farming continues in Chilika, for although the Tatas have withdrawn completely, there are many more small Tatas still there.

Orissa: mass movements against industries and mines

Save Gandhamardan Movement[11]
The first major popular struggle against mining in Orissa took place in the early 1980s, when a project run by the Bharat Aluminium Company (BALCO), a large public-sector organisation, was set up at Gandhamardan, in west Orissa. The area is rich in bio-diversity and is the home to countless threatened or endangered plant and animal species. It is also a place of great spiritual significance. In addition, Gandhamardan provides a life-support

system for about 100,000 people. The forests are especially important for marginal farmers and the landless, who account for roughly 30 per cent of the population, providing them with resources for their very survival.

When in 1971 the government of Orissa announced that Gandhamardan had a staggering 2–3 million tonnes of easily exploitable high-quality bauxite, the aluminium industry set its sight on these hills. However, it was not until 1981 that BALCO was given a lease on 36 sq km of forests to start opening up a mine. Initially, the local people were convinced by the government's claim that mining in Gandhamardan would bring prosperity and jobs to the area. But a visionary named Prasanna Kumar Sahoo, an administrative assistant in Sambalpur University, foresaw the damage that mining would cause. On his own, he began a crusade to make people aware of the hazards of mining in the area and the need to oppose it. Initially it was an uphill task, but his persistence and sincerity of purpose eventually paid off.

Among his first converts was Professor A. B. Mishra of the Life Sciences Department of Sambalpur University, who in time was to become a vital cog in the Gandhamardan movement. Under the auspices of the National Service Scheme at the University, he recruited student volunteers from various colleges, and gave them a 15-day training to enable them to run awareness camps in the villages. Some of the volunteers were so motivated that they chose to remain in the area and organise the people against BALCO.

The efforts of these committed people soon bore fruit. When the Chief Minister of Orissa went to lay the foundation stone for BALCO in May 1983, a mob stoned his motorcade, damaging vehicles and injuring government officials, an action that precipitated a violent response from the police. Meanwhile, public-interest litigation filed in the Orissa High Court had ensured that work on the project could not start till February 1985.

Frustrated by the non-cooperation of local people, BALCO attempted to recruit labourers from neighbouring States, but even they were persuaded by the agitators to return home. The government's approach to the problem was 'carrot and stick', promising largesse to those villages ready to co-operate, and threatening to withdraw support for village development projects from those which did not. BALCO also tried to buy support by doling out large donations to local youth clubs. But none of this worked.

When BALCO was finally allowed to start some preparatory excavation, the very first blast of explosives damaged parts of the holy Nrusinghanath temple. The perceived desecration only served to stiffen people's opposition. In February 1986 Sunderlal Bahuguna toured the area on foot for five days, and afterwards wrote a moving article in the *Indian Express*, which opened a national debate on the issue. Within a short while, journalists, environmental activists, and researchers converged on Gandhamardan, and the issue made national headlines.

The furore over Gandhamardan led to serious rethinking at the government level. Eventually after several environmental-impact assessments and many more protests, the government recognised that it had no option but to close down the project altogether, with effect from September 1989. When a new State government sought to revive it in 1992, albeit with a supposedly reduced environmental cost, more than 100 young protesters went on an eight-day cycle-rally-cum-protest from Nrusinghanath to Bhubaneswar and handed over a ten-page memorandum against the proposal to the Chief Minister. Under this pressure, he relented and shelved the revised proposal as well. But with such vast deposits of bauxite still untapped, the issue will not go away. Indeed in the last two years it has frequently hit the headlines, as proposals have been made for the exploitation of bauxite reserves and the establishment of alumina plants in Rayagada and Kalahandi Districts – proposals which have been challenged at every turn by local communities and activist organisations.

The Gopalpur steel plant[12]

In August 1995 the Tatas, who featured in the Chilika story, initiated the country's largest steel plant at Gopalpur, on the southern coast of Orissa. The capital outlay was Rs.93 billion, and the annual production capacity was to be 2.5 million tonnes. The project envisaged the acquisition of 5000 ha of government-owned and private land, displacing 30,000 people living in 25 villages. The people rejected the compensation package offered by the Tatas, because the area is economically developed, growing many high-value crops. The Tatas intended to pay Rs.100,000 per acre of land, as against an annual income of about Rs.50,000 per acre. The local people in the name of *Gana Sangram Samiti* (GSS: Mass Movement Committee) have formed a popular organisation, which has received support from various political and activist organisations, and NGOs.

Within a relatively short time, GSS succeeded in completely isolating the affected area, by banning the entry of both government and Tatas people. The agitators also blockaded the State Assembly in Bhubaneswar and sporadically blocked a major national highway and various district offices. Unable to suppress all this protest, the government arrested the leader of the movement under the National Security Act in July 1996. At the time of writing, the project has been stalled by the Tatas, citing the reason as a depression in the international steel market, and the government's lack of progress in land acquisition and infrastructural development.

Managing Orissa's forest resources: a people's movement

Orissa is a pioneer in community-based forest management, with a history that goes back several decades. In response to widespread deforestation, many tribal and non-tribal communities have actively taken up forest-protection work to a much greater degree than in most other Indian States. While some of these activities date back to the 1950s, and one even to the 1930s, most of it began in the 1970s. At a time when the Forest Department still regarded forest-dependent peoples as destroyers of forests and banned them from entering Reserved Forests, people started to protect a number of forest blocks, regardless of their legal status and the attitude of local foresters.

By the late 1980s, an estimated 3000 to 4000 communities in Orissa had established control over about 10 per cent of the State's forests, Reserved, Undemarcated, and Demarcated, covering some 572,000 ha.[13] Now the figure is close to 10,000. The spread of these is State-wide, but there are particular concentrations in Mayurbhanj, Bolangir, Dhenkanal, and Sambalpur Districts. There is a wide range of organisational structures, some of which have been formalised, and others of which have not. In some cases protection is informally overseen by groups of village elders or village forest-protection committees; in others by formally constituted voluntary organisations. In some cases the motivating factor has been the local presence of a voluntary organisation; in other cases a voluntary organisation has arisen from a local community-based system, as with BOJBP.

Summary and conclusion

In this chapter we have described a number of popular movements which have challenged the destructive aspects of modern development projects. We have seen how people have organised themselves to protect their livelihoods and their environment in a variety of ways. Most of these stories show some success in checking the onward march of 'progress' at the cost of poor and relatively powerless communities. In all cases the stories have revealed the blind disregard of governments and public and private-sector corporations for the lives of project-affected peoples and their rights to a livelihood. Space does not permit the telling of many more stories, both of communities who have been crushed (as at Hirakud) and communities who have been successful (as with *Chipko*).

What can we learn from these stories? First: the amazing strength and resourcefulness which people have shown in response to threats to their environment and livelihoods. Their stamina and courage in the face of intimidation and physical attacks in many cases are most impressive – all the more so when one considers that many of these people are from the most oppressed and intimidated sections of society.

Second, the diversity of organisations involved. While the creation of most movements has been in direct response to specific threats, many organisations and individuals have brought particular experience, perspectives, and conceptual thinking which have clarified the analyses of the activists and strengthened their strategies. Many of the early activists were from the Gandhian tradition, believing in the non-violent model of development against capitalist developments. People like Sunderlal Bahuguna represent this approach. It may be said that Bahuguna's environmental perspective has always been that of the pure conservationist: '*No human being can live without nature. Therefore nature must be conserved to protect human beings.*' Others, of a radical political persuasion, such as USV in the Himalayan hills, argue: '*Nature cannot exist without social justice. When social justice is achieved, nature will be secure.*' They believe that the environment cannot be protected without first addressing the issue of the production and distribution of resources within society. USV's environmentalism, therefore, is strongly focused on rights and justice. Between these two is a range of positions, which in different ways aim to strike some sort of balance between pure conservation on the one hand and human development with equity on the other.

Third is the issue of tactics. The Gandhians have always eschewed violence, using the tools and vocabulary of the freedom movement to succeed: fasts, mass rallies, and *padayatras*. But more recently some movements have adopted insurgency activities to disrupt development projects, resorting to demonstrations, *gheraos* (blockades), *hartal* (disruption of economic life), jail *bharo* ('fill-up jail'), and court action (particularly public-interest litigation) to achieve their objectives and demonstrate their power to influence the government on environmental and livelihood issues.

Now the network of environmental activists, environmental organisations, NGOs, and people's organisations has become so strong that not a single project that might adversely affect the environment and the livelihood rights of people is being approved without their knowledge.

BOJBP: from birth to maturity 5

'We forgot our painful efforts of times past by looking at the greenery created by us.'
(Gopal Pattanaik, retired teacher at Awasthapada High School, and a resident of Manapur village)

The movement takes shape

Chapter 1 described the early years of BOJBP: years in which, in the words of Joginath Sahoo, the movement was *'without shape or size and also name'*, struggling to give birth to itself. 1982 was the year in which that finally happened, when all 22 villages, clustered around the two hills, united in the common cause of forest regeneration and environmental protection.

Internal challenges and external threats

For all the movement's early achievements, it would be a mistake to suggest that the years of struggle were over, for as one set of problems was overcome, so a new set arose to take its place. For example, as the forest started to regenerate, wildlife started to return: deer, bears, rabbits, monkeys, snakes, and a great variety of birds. With the return of animals, inevitably came the prospect of a return to hunting. Throughout India, for communities living in or near forests, game provides a valuable supplement to meals that are usually meat-free. In this respect Kesharpur and neighbouring communities were no exception. But to Joginath Sahoo – if not to many others – the protection of wildlife has always been as important as the protection of trees: not for nothing was the movement called Friends of Trees *and* Living Beings. On more than one occasion, therefore, he had to use all his considerable powers of persuasion to dissuade villagers from giving chase to game.

In addition to these internal challenges, there were considerable threats from outside. Just because the villagers of the 22 villages had united to protect their environment, it did not mean that external players would respect their values and their actions. For example, a contractor turned up in April 1983 with his men to quarry stone from the hill above Binjhagiri village, on the other side of the hill from Kesharpur. He needed it to build a bridge over a nearby river. But he was confronted by an action that he could not possibly have anticipated. All 30 families of this tiny village ran to the spot where he and his men were working and, in the words of the Hazaris,[1] 'offered civil resistance'. The exact form of this resistance is not described, but it is likely to have been non-violent and inspired by Gandhian practice. Whatever it was, it achieved its purpose, and the contractor backed down: 'the forces of destruction', as the Hazaris call them, were vanquished, at least temporarily.

Fighting inertia and cynicism

Coupled with the need for vigilance against external threats, there was also the need to deal with internal inertia and cynicism, which presented a challenge of a different kind. Manoj Hazari, a youth activist from Kesharpur, told us that in the early years there were those who mocked the movement with comments like: '*Why should we protect this denuded hillock? What would we get from it?*' The leaders' response to such negativity is illustrated in this interesting story told by the Hazari brothers. In July 1983, the National Service Scheme (NSS) Unit from Sarankul College failed to arrive for a session of tree planting on Malatigiri. Frustrated but not thwarted, Jogi went from village to village and door to door, begging people to come forward to help out. But his entreaties were largely unsuccessful: he was able to persuade only a few school children to join him. It was late by the time they started, and the work turned out to be extremely arduous, owing to the very hot weather and the fact that the ground was hard because the monsoons had failed. But then, after a day's exhausting labour, almost miraculously the heavens opened, and down poured life-giving rain: the monsoons had arrived. According to the Hazaris, this had 'an electrifying effect on the entire area' and resulted in men and women coming forward from many villages to help in the plantation work, not just the next day but over several days following. This was the first time that women had volunteered to participate in 'this sacred endeavour', which the Hazaris describe as 'a sight for the Gods to see'. What had started so inauspiciously turned out to be an amazing success. Inertia was challenged, the Gods were propitious, and the job was done!

It is interesting to note the language that the Hazaris use: 'forces of destruction', 'sacred endeavour', 'a sight for the Gods to see'. From its very beginnings, at least among the core activists, there was a very strong religious and moral dimension to their work, which gave it an imperative beyond pure self-interest. Many people who made their way to Kesharpur came (and still come) as pilgrims visiting a distant temple on a moral and spiritual quest. On his visit in late 1983, Sunderlal Bahuguna,[2] the renowned leader of the *Chipko* Movement, spoke of Kesharpur as a 'centre of pilgrimage for all who are working for the survival of the living beings on this dying planet'.

Engaging with Oxfam

In the middle of 1984, some two and a half years after the birth of the movement, BOJBP made its first approach to Oxfam for funding. Fire had swept through Awasthapada, a village on the other side of the hill from Kesharpur, severely damaging 600 houses, including some weavers' cottages, together with their looms. Virtually all houses in the villages of this area are thatched, and fire is an ever-present hazard. Since houses are built very close to each other, once fire breaks out it rapidly spreads. In the dry season in Orissa it is reckoned that there are at least two major fires and five minor fires every day. Some districts are particularly susceptible, because of the preponderance of thatch; Nayagarh is one of them. Oxfam's first grant to BOJBP was an emergency payment of £684, which enabled five very needy families to rebuild their houses, and the handloom weavers to buy replacement equipment.

Having made contact, BOJBP approached Oxfam later in the same year, seeking for the first time support for their environmental work. By that time the movement had spread its message from the 22 core villages to a further 30, with plans to work with 100 more in the near future. Visiting the area after the application was received, Oxfam's Project Officer commented on the extent to which the forests had visibly regenerated, on the return of wildlife, and on the regeneration of springs. He also noted the existence of the 'Green Club' in the village of Manapur at the foot of Malatigiri; the involvement of teachers and children in the movement; and the very democratic process by which the people in Kesharpur chose their village council, of which more later. Oxfam's made its first development grant: £1235, with which to buy audio-visual aids, a typewriter, a duplicator, and two cycles, and to pay the salary of one worker to help to co-ordinate activities and formalise administrative systems. With its plans to spread its message into many more villages, the movement needed more robust systems and procedures, both to consolidate and to expand its work.

Spreading the message

On the march

Chapter 1 referred to some of the awareness-raising techniques used by the BOJBP movement. We now consider in more detail how these techniques have been used. In its first ten years, one of the most effective and enduring tactic was the *padayatra*, the campaigning march. Thus in February 1984, as part of their campaign to reach more villages, a *padayatra* involving 128 volunteers from four villages set out from the village of Chinara, on the road between Kesharpur and Nayagarh. In three days they covered 35 villages, spreading the message of environmental protection. The influence of this event on the communities they visited is described by the Hazaris: 'In village after village the *padayatris* were received with warmth and affection'. In only one were they rebuffed because there were *dalits* (former 'untouchables') in the group.

In subsequent years the movement's *padayatras* grew in size and coverage. In January 1987, for example, as part of a National Environment Awareness Campaign, they held a series of *padayatras* over a three-week period, covering no fewer than 100 villages in six blocks[3] of what was then Nayagarh sub-district. In the course of this major event, 'hundreds and hundreds' of villagers were 'awakened' for the first time to the need for environmental awareness, 'as it was a new idea for them'. Again notice the choice of language in which these events are described in BOJBP reports, which is reminiscent of that used by religious preachers and missionaries. Perhaps we should not be surprised, for the core activists saw themselves as 'environmental missionaries' with a duty to 'spread the word' and call on people to make a commitment to the environment.

Over the years the numbers of *padayatras* has been countless, and their importance cannot be overstated. Dr Hazari told us recently that he has more faith in *padayatras* than in workshops. '*You have to suffer for the people*', he said. '*Otherwise people won't follow your suggestion.*' He went on, '*If one person goes by a vehicle and motivates the people and the other person goes on foot ... then definitely people would listen to the latter.*'

Occasions to celebrate

Padayatras – large and small – were frequently used as part of the celebrations of both national and global events. For example, on 5 and 6 June 1987, BOJBP celebrated World Environment Day, with a two-day *padayatra* passing through villages at the foot of Ratnamalla Mountain. On the way they held short 'wayside meetings' in every village to discuss forest protection and plantation, wildlife protection, and the cause of forest fires. This occasion was also used to send no fewer than 10,000 letters to school children in the area, asking them to participate in the next event in their calendar of activities, namely a tree-planting programme that was planned for *Banamahotsava* ('Forest Festival'), which was celebrated in July.

Perhaps more than any other celebration, *Banamahotsava* was a particularly significant festival in the early years of the movement, when much tree planting was done and when relationships with key external players were strengthened. At *Banamahotsava* in 1985, for example, the former and current Chief Conservators of Forests in Orissa were invited to join in activities, in order to reinforce BOJBP's relationship with the State Forest Department. Then in 1989 a public meeting was organised in the tribal village of Kusapanderi, to which again senior foresters were invited. On this occasion, after a visit to the village's 40-acre protected forest, a presentation was held, at which the community were awarded a certificate and a shield in recognition of their work of forest protection since 1980. Sweets were presented to the children.

Thinking globally, acting locally
Festivals and special 'days' and 'weeks' were not simply occasions for celebration, but were cleverly used as a means to an end, that of spreading the message of environmental protection and other messages. The movement has always sought to 'act locally', while 'thinking globally'. Thus Wildlife Conservation Week, a global event held in October every year, was used in 1988 to form the central committee of a new 'sister organisation' to protect Balaram Mountain close to Nayagarh. Human Rights Day in December has frequently been used to campaign against dowry and untouchability.[4] ('The natural environment is polluted, but ... the social environment has also been spoiled due to ... untouchability and dowry', says a BOJBP Report.) World Health Day in April 1987 was used to raise awareness about hookworm, which 'is more dangerous than the tiger'. In January 1990 a number of events during Leprosy Eradication Fortnight were used to raise awareness of this disease, which is very common in rural Orissa, and which has always been of concern to the movement. On World Forest Day in March 1987, leaders organised a mass meeting for people from 25 villages in Khandapara Block, at which the atmosphere distinctly warmed when some hard questions were asked of the forest officials who were present. This event ended with a cultural evening: a drama about forest protection and plantation, at which the need 'to act locally and think globally' was reinforced.

Songs, slogans, dramas, and publications
'Cultural groups are the best media for propaganda and mass mobilisation on environmental conservation and awareness,' says the BOJBP Work Report for 1991–92. To this end a huge repertoire of songs, slogans, poems, and dramas has been developed, cleverly based on traditional idioms and traditions. Here, for example, is a verse from one of their songs:

> *Brikshyo' aamaro jeevan dhano,*
> *Jagaye mati pani pawano,*

Brikshyo' bina jeewan nahi,
Brikshyo' aamaro jeevan bhai.

Trees are our life's treasure,
They offer soil, water, and air,
Without trees life would end,
Trees are our life's friends.

Are these songs traditional or modern? In May 1997, while on a visit to Britain, Jogi sang a song much like this one to an audience in Oxford. When asked whether it was traditional or if he wrote it himself, he replied, '*It is a traditional song that I wrote myself!*' According to Ajay Mohanty, youth leader from Manapur, Jogi's brother, Bhagaban, has also been a great 'composer' of songs and slogans and was much praised by Sunderlal Bahuguna on his visit to Kesharpur in 1983.

In the same way, he and others have devised many dramas, particularly for children and young people to perform. In January 1990, for example, 1200 students attended a special festival at Nuagaon. There was an exhibition of posters, and balloons were distributed with messages on them – about the need for clean air, clean water, economic use of fuel, protection of wildlife, and an end to dowry, untouchability, and 'traffic disturbances'.[5] The event included a fancy-dress *padayatra*, followed by a drama, specially written for the occasion, and a film about wildlife.

In 1992 BOJBP trained a special cultural team which went to 15 villages to spread the word on forest conservation through drama, songs, and slogans. In the same year they held a two-day convention, to which they invited poets, novelists, artists, and cultural groups to explore ways of using a variety of cultural forms to spread the message of environmental awareness.

In support of all their activities, over the years BOJBP has published huge numbers of leaflets, posters, and booklets, simply written in Oriya, the language of the State, conveying and reinforcing environmental messages to ordinary villagers. Every campaign begins with the circulation of leaflets to all households in the target area. These leaflets use catchy slogans and often carry a message from religious or moral literature. They are frequently written in the name of a tree, or hill, or village. Thus a tree will 'request' people to save her and her environment, and ask them to undertake planting more trees in her area. The booklets have ranged from practical manuals on health care, on planting and nursery raising, to collections of wise sayings. (A list of publications is given in Appendix 2.) A mail-order catalogue is published for the seed bank, updated every few years as the stock has grown.

(More details of the seed bank are given later in this chapter.)

Funerals, weddings, trees, and children

The movement has not confined itself simply to specially created cultural events. It has also very creatively used existing rituals as a means of spreading its 'green message'. On the death of someone who has been particularly active in the movement, the occasion has been used for relatives and friends to plant trees as a memorial to the deceased – and not just one or two: in 1992 the death of a particularly dedicated activist was marked by what was described as a 'mass plantation'.

At the other end of life, BOJBP has always encouraged parents to plant trees to mark the birth of a child in the family, and married couples to give one another saplings at their weddings, in place of a dowry (Figure 7). In 1991, people were called by special invitation to attend the wedding of a couple in Godipara village, at which trees given by the couple's parents were planted by the bride and groom in the grounds of the local health centre. In the evening Jogi gave a speech on 'the responsibility [of the new couple] towards family and society'. (Again, note how the event was used to convey a moral message.)

Figure 7: The poster shows a couple exchanging saplings at their wedding. BOJBP has campaigned against the dowry system because it is such a drain on the resources of the bride's family and can lead poor families into severe indebtedness. By encouraging the exchange of saplings as a substitute for dowry, not only are they challenging a harmful traditional practice, but they are also putting trees at the heart of an important ritual.

Commenting on this special occasion, the Annual Report for that year observes with some pride, 'Perhaps this type of invitation was printed for the first time in the world.'

All of these examples serve to demonstrate the ways in which the movement has aimed to create a green awareness through a green culture. Nowhere is this seen more strikingly than in schools, where, as soon as they can write, children are asked to send letters to trees – from whom they receive replies (written by their teachers!). Schools have their own tree nurseries and tree-planting events; children are encouraged to write and take part in their own dramas; and are urged to enter inter-school 'green' debating competitions, for which the prizes are 'green stipends': bursaries to cover school expenses. And while environmental studies is not part of the formal curriculum, nevertheless many teachers involved in the movement will use the opportunities offered in many subject areas to convey the message of environmental awareness, and the need for environmental action. The schools programme is a vital part of the work of BOJBP.

Tapping into a 'green spirituality'

Just above Kesharpur, within the forest, is a little spring, one of several on the hill, which feeds a tiny stream that flows into the Kusumi. In the days when the hill was totally bare, the spring had completely dried up, because the water-table had dropped so far. But as the forest regenerated, so the water-table rose and the spring started to flow again. When that happened, a small shrine to the goddess Durga was created at the site of the spring (Figure 8). Now, when people go there, they often kneel and make a *puja* (an act of prayerful devotion).

There is no doubt that Friends of Trees have tapped into a deep vein of 'green spirituality' that lies within rural Hinduism, a spirituality which embraces all living things. However, while recognising the significance of the spiritual–cultural dimension in their work in the 1980s and early 1990s, given the stresses that developed later in the movement, it is important not to overstate its longer-term significance among most ordinary villagers.

Sowing 'seeds' for others to grow

In the mid-1980s, a couple of years after the birth of the movement, a decision was taken to scale up activities and replicate good practice. In this BOJBP was faced with a choice: either to add more member villages to the core 22 villages of the 'mother area', or to set up parallel 'sister organisations', made up of clusters or federations of villages. They chose the latter course as the better way to facilitate local solutions to local problems and to develop local capacity

BOJBP: from birth to maturity

Figure 8: At this recharged spring in the forest above Kesharpur, villagers have made a shrine to the goddess Durga, where villagers often perform a '*puja*' – an act of religious devotion. There is, thus, an important spiritual dimension to the environmentalism of many of the core activists involved in the movement.

in the protection and regeneration of forests. In the district there are many hills, mostly covered, or formerly covered, with *Khesra* Forests (Undemarcated Protected Forests: see Chapter 2). Some of these are smaller than Binjhagiri, others are much larger, and some are quite mountainous, such as Sukarmalla and Sulia. All are the rugged stumps of ancient mountain ranges, almost worn away by time. Each required a cluster of villages to protect it, and the larger ones needed two or three. In each cluster there were up to 30 villages, and the structure of the sister organisation replicated that of BOJBP. Elsewhere, small patches of forest were covered by single, independent, village forest-protection committees (IVFPCs).

By late 1993, according to an Oxfam paper,[6] 18 sister organisations had been formed, covering 301 villages in six blocks of Nayagarh District. In addition there were 23 IVFPCs. The total area of degraded forest, which they covered, was 35,000 ha, spread over 18 hills in the District. Once these entities were established, BOJBP saw its role as one of support and motivation. For this it ran training workshops, camps, *padayatras*, and cultural events, using its own volunteers and those of the organisations themselves. The initial aim was to make the new organisations as self-supporting as possible, while staying within the BOJBP fold and drawing support from it. It was out of this network that a central committee for forest protection, the *Nayagarh Jangala Surakhya Mahasangha*, was formed in 1992 – about which more is written later in this chapter and the next.

Entreaties, prostrations, protests, and fasting

When confronted with people who were taking the narrow self-interested view of things or who were simply being obstructive and factional, Jogi and the other leaders had four tools of persuasion at their disposal.

First they would always seek to persuade people by entreating them and by employing reasoned argument. Jogi told us that in the village of Godipalli there had been a great deal of factionalism for years, which had paralysed communal activities to such an extent that the village pond became completely choked, because no one took responsibility for cleaning it out, and the temple was closed for 12 years. Taking with him two key BOJBP workers, Susanta Jena and Priya Nilimani, Jogi went to the village to talk through the problems with the villagers. They met at 9 at night. '*There were many hot arguments and counter-arguments,*' Jogi said, '*which went on for six hours.*' Finally at 3 in the morning, presumably in part because people were exhausted, there was a break-through, and reconciliation was reached. The villagers agreed to work together, and 17 young people came forward to manage the village affairs. Reason and good sense had prevailed.

In the history of the movement there have been literally hundreds of occasions like this: all-night meetings to settle disputes between villagers and villages. What they show is the extraordinary tenacity and stamina of Jogi and his colleagues, and their commitment to using reason and entreaty to reach understanding, reconciliation, and agreement.

Touching gestures

If reason and entreaty failed – and frequently they did, when passions were aroused and entrenched positions taken – BOJBP workers would use the technique of prostration: falling at someone's feet. In Indian culture the power of this gesture is immense, having both an emotional and a moral dimension. If you touch someone's feet, or prostrate yourself before them, you are showing them that you respect them, and that you do not consider yourself to be above them. In families, children sometimes touch the feet of their parents and grandparents to show them respect. Outside the family you may also touch the feet of anyone whom you regard as being of a higher status than you and worthy of your respect. By prostrating himself in the many confrontations in which he was involved, Jogi was showing that even he, as a respected person, did not consider himself to be so important that he could not give respect, even though he disagreed with the other person's position. Although it did not always work, very frequently it did.

Thus, to tell another hunting story, when confronting the villagers of Puania who were intent on killing a deer that had come out of the forest on to their cropland, Jogi and two friends at first entreated them to spare its life. But this did not work: in fact the hunters were outraged to be asked. So the three started to chant songs and slogans, exhorting the villagers to think of the importance of animals in the local eco-system. After a while the children of the village joined in, which made the hunters even angrier. At that point, with tensions very high, the three workers fell at the feet of the villagers and begged them to spare the life of the innocent animal. At that point, write the Hazaris, 'hearts melted and the prayer was granted'.

Sit-ins and fasts

On other occasions, peaceful 'sit-ins' have been used to powerful effect. Banchhanidhi Pradhan of Gamein told us this story of how a conflict was resolved through a peaceful protest:

> Once I had been to Sanagarada village, along with 200 members of the organisation, including women, to resolve conflict with the village. We all slept on the village road from 4 o'clock in the morning till 11 am. We did not take any food. Some of our members were moving from house to house to convince them to resolve their internal conflict. Finally the leaders of the various factions agreed to sit together and discuss their problem. In the evening a village meeting was convened, and all the factions in the village united and resolved to stay together without any tension and group fighting. This was a very good step for resolving village conflict.

Alaka Nanda, Coordinator of BOJPB's Women's Development Programme, told us how she and a colleague, Bhagya, by themselves held a similar protest in Puania to confront people there who had illegally felled trees. As a result, *'All the villagers deposited their timber before the village, and it was auctioned.'* The money so raised, Rs. 30,000 (a substantial amount, more than £450), was then deposited in the village fund, so that no individual would profit from it. Clearly weight of numbers was not always necessary.

Many other instances of the use of prostrations and peaceful sit-ins are described by the Hazaris and by villagers whom we interviewed. Also described are fasts, which have been used when all else failed. In 1981, before the formation of the movement, Narayan Hazari was concerned about the half-hearted response of many villagers to his efforts to persuade them to take environmental conservation seriously. 'Convinced that my words would not move them,' he writes, 'I decided to fast.' He fasted in seven villages over the period of a week. Jogi too has fasted 'more times than he can remember'. In conversation with us he described one such occasion:

> One day when school finished early – late morning – I decided to walk home to Manapur [his village, just up the road from Kesharpur] through the forest. On the way I noticed that someone had felled a tree, carelessly leaving the loppings behind. I realised from the proximity to the village of Nagamundali that most likely it was someone from there who had done it. So I picked up the discarded branches and went with them into Nagamundali, where I sat on the road and fasted. After several hours people gathered round and asked me what was going on. I explained the purpose of my fast. Immediately they knew who the culprit was and subsequently reprimanded him and fined him Rs. 500.00.[7]

More forceful forms of persuasion

Occasionally, just occasionally, villagers – if not their leaders – resorted to more forceful means of persuasion. Bankanidhi Mohanty of Puania told us how once a group from his village seized a tractor loaded with stones, taken from the nearby hillside. They impounded it until the contractor and his men *'after a long negotiation ... were convinced [!] to stop stone quarrying in the area'*. A rather more violent incident was related to us by a teacher from the Middle School in Kesharpur, Baikuntha Pattanaik:

> Some of the villagers of Kesharpur were cultivating *ragi* and other millets on some portion of Binjhagiri, and that area belonged to Gambhardihi. Leaders of Gambhardihi informed the leaders of Kesharpur. But the leaders could not stop the cultivation. The youths of Gambhardihi are highly organised and they destroyed the crop. After that the villagers of Kesharpur have never undertaken farming on the hill.

It must be stressed that this was a unique action, taken out of frustration and without sanction because the people of Gambhardihi could not get what they needed, which was an effective intervention from BOJBP.

Seeds and saplings

A year after its first development intervention, in May 1986, Oxfam made a two-year grant to BOJBP: £2095, to cover the costs of meetings to promote environmental awareness and protection, to publish a booklet, and to meet core running expenses, which included one and a half salaries. Six months later, a second two-year grant of £1885 was made to set up and run a seed bank. This initiative was designed in consultation with the Orissa Department of the Environment, with funding from Oxfam, and it was one of five seed banks that were to be set up by Oxfam partners in Orissa. Their stated purpose was 'to promote the cultivation of a variety of local species of trees so as to bring financial and environmental benefits to local communities'. Oxfam's Project Report, 'Preserving Ecological Diversity in Orissa', describes the idea behind the seed banks:

> Seeds of indigenous tree species which are known to be useful will be collected in two ways. Each agency will employ two full-time collectors, who will be kept busy all year round, since different species seed at different times. The agencies will also buy seeds from local people. The seeds will then be sorted, fumigated and stored, ready to be sold to local groups, schools, and individuals at prices intended to encourage the use of these seeds.
> The Department of the Environment plans to encourage tree nurseries at schools in the relevant areas. These should provide a steady market for the seed banks. Once the seeds are sown and germinated, they should provide an income for schools and groups when they are sold and planted out.

It was an excellent idea, enthusiastically taken up, but one that proved difficult to turn into a fully functioning reality. First, for various reasons, the other four organisations never got their seed banks up and running. Second, even with BOJBP there were problems, for gathering seeds proved easier than marketing them. Insufficient numbers of schools and other local groups established nurseries to take seeds, and insufficient orders came from outside agencies. As a result, in their first year the collectors gathered seeds from 37 varieties, but managed to sell only a quarter of their stock: a very disappointing take-up. This meant that at the end of the season they had a vast surplus of seeds, which would be wasted since their germination could not be assured beyond one year. Just when they were thinking of throwing them away, they received a visit from Professor Radhamohan, an old friend, who persuaded them to sow them: first, in a designated plantation area on the hill, and second in their own purpose-made tree nursery. From this nursery they could then distribute seedlings to people in the communities around. But to do this they needed land, and to purchase that land they needed funds. So they launched a public appeal and bought just over 0.1 ha (0.25 ha) on the edge of the village, on which to establish the nursery, and a little later an office.

To meet the expected demand, it was decided that the nursery needed to maintain a stock of 10,000 seedlings, and that these should be categorised by

Figure 9: The nursery behind BOJBP's office in Kesharpur was founded in 1987, when there was a surplus of seeds in the seed bank and members did not want to throw them away. Professor Radhamohan suggested they should sow them in their own nursery.

their utility: medicinal, aesthetic, or economic (as fodder, fuelwood, timber, etc). By the end of 1988 they had achieved that level from their own seeds, with 31 varieties being grown. Figure 9 shows a more recent view of the nursery.

While the size of the stock remained fairly constant, over the next few years its range expanded considerably. By 1993, which was the year of a comprehensive joint Oxfam–ODA[8] evaluation, the nursery had more than 100 species in stock, of which 61 species were distributed, 95 per cent of these being indigenous. The authors of the ODA Report note that BOJBP told them that the volume of 'sales' had increased year on year. ('Sales' refers to everything that was disposed of, both through actual sales and through free distribution.) They also reported an average of 580 families per year taking plants for their back-yards and farmlands, for which they were charged Rs.0.5 per sapling. (People taking trees for communal planting received theirs free of charge.)

The seed bank has never been as successful as hoped, although sales have increased year on year. Since the early 1990s, the mail-order catalogue has offered seeds of more than 100 species, almost entirely indigenous. The 1994 ODA Report notes that 77 per cent of seeds supplied went to organisations in Orissa, with most of the rest going to other States, and a fraction overseas.

Take-up by government bodies has never been as great as hoped, for the basic reason that BOJBP's aim has been to promote indigenous species, while government agencies are interested almost entirely in exotic varieties, such as eucalyptus. The trouble has always been that, while the organisation has used mailings and newspaper advertisements to promote the seed bank, it has never had a proper marketing strategy. There have also been practical problems of collection. (BOJBP have not used sister organisations to help them – although they could have done.) There has also been a tendency to keep the seeds too long, leading to poor germination rates.

Getting gender on to the agenda

It was not until 1986 that BOJBP, on paper at least, acknowledged the importance of involving women in the work of the movement. At the end of their Annual Report for that year, they outlined their programme for the next two years, and included a 'women's workshop on environmental education and alternative energy'. Where and when this took place is not recorded. But two years later, Oxfam's Project Officer, Jagdish Pradhan, commented in a report, 'So far BOJBP has been taking up forest protection activities without putting much emphasis on the poorest ones and women.'

In spite of pressure (or rather perhaps because of inadequate pressure) from Oxfam, women continued to be excluded from the processes of environmental protection. However, at the end of the 1980s and in the early 1990s BOJBP organised training programmes for women, promoting low-cost latrines and smokeless *chullas* (cooking stoves), teaching tailoring and embroidery ('for poor women and girls'), and holding classes on environmental education, sanitation, family health, and family planning. A year later, more training focused on issues related to women's work and health, but did nothing significantly either to develop their role or to address their position within the organisation or in wider society.

At last in their three-year budget proposal to Oxfam for 1992/93, BOJBP acknowledged: 'Women's participation in the environmental conservation programme launched by BOJBP is poor.' But what they proposed was basically more of the same: training in health and family welfare, environmental education, and tailoring, for which they proposed the appointment of a core of women workers, although under the management of a man. However, between the proposal being drawn up and the grant being made, Oxfam insisted for the first time that there should be a gender-training workshop for BOJBP staff. This happened at the end of 1992, facilitated by Oxfam staff. Women's groups were set up in 15 villages, and an exposure visit was organised for ten women to another Oxfam-funded organisation in Orissa, which helped to increase their confidence and understanding.

So at last, ten years after the formation of BOJBP, gender issues started to be addressed seriously. But just how seriously was questioned by Mamata, Geeta, and Satyabhama, three women workers with BOJBP who told us that things are not what they might appear: *'Women's organisations have been created, but they are not properly integrated into the mainstream forest-protection/environment-conservation movement.'* We shall return to this issue in Chapter 6.

System of governance

Before considering further developments in the 1980s and early 1990s, it is important to look at the system of governance that the movement developed in its early years. Governance operated on two levels: at the village level and at the 'village federation' level, that is the level of BOJBP itself. All 22 villages in the 'mother area' of BOJBP had long had traditional informal Village Councils, responsible for managing a number of village resources. These included schools, temples, village common lands, and village ponds. The size of these Councils varied from five to ten members, and included in most cases a president, secretary, and treasurer as office bearers.

In most villages, members of the Council were traditionally selected by a General Body of the village, which itself was made up of one member representing each household. Very early on, Kesharpur evolved a different and very democratic system, which reduced the possibility of nepotism. In their system, every adult over 18 years of age would cast his or her vote by secret ballot, in which there were no nominations but in which the voter was required to write down the names of five people of his or her choosing. The five people whose names occurred the maximum number of times would then be asked to serve on the Council.

In all villages any member could be removed from the Council if the villagers lost faith in him or her. In addition to its responsibilities in the management of common-property resources, each Council was also required to maintain accounts, which were presented annually to the General Body of the village, or in the case of Kesharpur to the whole village.

Rules and regulations framed by the villages

Very early on, in each of the nine communities that lie at the foot of Binjaghiri, the Village Councils framed sets of rules through which the rights and responsibilities of the villagers over the forest were defined. This set has always been more or less common to all nine villages.[9] The rules are as follows:

1. The forest is to be protected by voluntary patrolling on a rotational basis, using the system of *thengapalli* [described in Chapter 1].
2. Every household is to participate in *thengapalli*, but in circumstances where default is unavoidable, exchanges of duty are allowed. Refraining from this duty purposely without informing or without adequate reason invites compensatory duty on two days instead of one.
3. No one can cut a tree from the forest, except in emergency, but even then only with the permission of the Council.
4. Villagers can collect dry fallen twigs, fruits, seeds, and flowers and can cut some shrubs like *pokasunga* which have been identified by the Council for fuelwood.
5. Protected areas are closed for grazing until natural regeneration or plantation is established.
6. Nobody can enter the forest with an axe, without prior permission.
7. Villagers can collect stones for construction from the forest area for *bona fide* use only, for example for the construction of cooking stoves.
8. In the event of any threat to the forest from outsiders, every villager must help the one who is on *thengapalli* patrol.
9. Those who are caught violating rules can be fined, the level of fine being decided by the Council. [Normally, however, a public apology would be called for. Alternatively the culprit would be asked to plant trees and care for them until they were well established.]

To this list Kesharpur added another rule concerning trees along the bank of the river Kusumi. The Council decided that river-bank trees should be cared for by the farmers who owned adjacent fields. When the time came for them to be felled (which the Council would decide), the wood would be divided equally between the guardian farmer and the rest of the village, as we described in Chapter 1.

While these rules operated in the nine villages at the immediate hill-foot of Binjhagiri, the other 13 villages agreed to restrain themselves and refrain from damaging the environment on the hill. (Of these 13, seven initiated protection of other forest patches in their vicinity, of which the Malati forest was one.)

The General Body

To deal with governance at the supra-village level, a General Body of 101 members was formed, consisting of representatives of all 22 villages, with places allocated in proportion to the sizes of their populations. This General Body, which met three or four times a year, was responsible for electing from among their numbers an Executive Committee, which was in turn responsible for all major policy decisions relating to the organisation. It was decided that the Secretary of the Executive Committee should be its chief functionary, responsible for implementing the decisions taken by the Committee. But, of course, the Secretary could not carry out those decisions alone, especially since he (and it always has been a 'he') was a volunteer. To support him, therefore, there was a core of key volunteers, some two dozen or so in number, and a few paid staff. With individual Village Councils being responsible for protecting their adjacent hillsides, the Executive Committee of BOJBP was responsible for developing all the programmes of awareness raising, for the establishment and training of sister organisations, for the seed bank and tree nursery, and for managing the paid staff.

For many years, paid staff numbered between one and three, but by the early 1990s the work demanded more, and in 1992–93 the numbers were increased to 11. This increase changed the dynamics between staff and key volunteers, and several difficulties arose as staff sought to increase their influence and to break out from the subordinate position that they had held relative to the volunteers. This issue is considered again in the next chapter.

Growth, and the price of growth

In April 1988, Oxfam received from BOJBP a proposal for a very ambitious two-year programme of work. A huge range of activities was proposed, which was a reflection of both the functional increase in their work and its increasing geographical spread. 'During the last three years,' says the proposal, 'the *Parishad* has extended its movement through various programmes to 400 villages.' This does not mean that the 'mother area' of BOJBP had expanded – it has always remained as the core 22 villages – but the area of 'spreading the message' and supporting sister organisations had grown enormously. What follows is a list of the activities planned for the years 1988/89 and 1989/90, for which they required a grant from Oxfam:

1. Organisational meetings, both within the villages of the mother area and in the villages of sister organisations, covering environmental conservation, the cultivation of green vegetables, health and sanitation,

leprosy eradication, and voluntary labour. On the latter subject, the proposal says: 'As it is seen that people are becoming self-centred day by day, it is necessary to beat the drum again and again for social and community development.'

2. Both within the mother area and beyond, celebration of World Environment Day, *Banamahotsava*, Wildlife Conservation Week, Human Rights Day, and World Forest Day.

3. Promoting organic farming: educating farmers on the dangers of agro-chemicals and helping to wean them off them.

4. An essay competition, an exhibition, and the implementation of 'green stipends': bursaries to cover expenses for school books and uniforms, which were won in inter-school debating competitions.

5. Three postcard campaigns to raise people's awareness on environmental issues in new villages.

6. Exposure visit for core members of the movement to see and learn from the work of like-minded organisations.

7. Hospitality for visitors, of whom there were increasing numbers.

8. The supply of experimental low-cost earthen latrines. Activists in the movement had been concerned for some time about the health risks associated with uncontrolled defecation. They wanted to encourage villagers to construct pit latrines, for which BOJBP was prepared to supply the pans at a subsidised rate.

9. Publication of a book in Oriya on the technical aspects of nursery raising, plantation, and forest management.

10. Establishment of a small library of newspaper cuttings, newsletters, and cassettes.

11. Training programme for a small environmental education team in each village of the mother area.

12. Office building. Until 1988, BOJBP had operated from three small rented rooms in Kesharpur. They had already started to construct a three-roomed building on the land that they had purchased with public donations, where the nursery was now established. They needed funds for its completion. Figure 10 shows the completed office.

13. Running costs of two motor cycles which had been purchased with an earlier grant and which were needed to cover the wide area of contact villages now involved in the movement.

Oxfam's response

Before the grant was made, Oxfam's Project Officer, Jagdish Pradhan, visited BOJBP. In his tour report, dated 26 May 1988, he commented on several aspects of their work, including their growth: 'During 1986–88 BOJBP has been able to establish direct contact in 290 villages and indirectly has influenced more than 400 villages in Nayagarh sub-division. In all these villages people have been motivated to protect their village forests, take up plantation and organise themselves.'

As we have seen, the way in which they had done this was to set up sister organisations of village-level groups as well as independent village forest-protection committees (IVFPCs). By 1993 this had brought more than 320 villages into the BOJBP fold – an astonishing achievement of organisation and motivation. However, Jagdish noted that growth had come at a price:

> [BOJBP] are now under pressure of time, as the number of contact villages has become very large, but they are not interested to recruit [more] paid staff for supporting them. I feel that BOJBP will have to rethink its management strategy and programmes during the next one year. Otherwise it may become less effective, and the newly organised village groups may disintegrate without getting timely encouragement from BOJBP.

Figure 10: BOJBP's first offices were in rented rooms in the village. This plot of land, on the edge of the village, was bought in the mid-1980s by public subscription, and the movement's headquarters were built on it. It comprises a meeting room, seed-bank store, and administration office. Since then, with Oxfam support, another small block has been built to the right of the main building, housing two guest rooms and a garage-cum-store.

Also attracting critical comment was the organisation's failure to relate its work to the problems of deforestation at the State level; its failure to involve women in any significant way in the movement's activities; and its failure to consider the impact of its activities on the poorest families and upon women. Jagdish was also critical of its failure to create a 'cadre of activists' to carry out organisational work in each village, which led to huge over-dependence on the small core of central activists grouped around Jogi, Udayanath Katei, and Narayan Hazari. This over-dependence became a critical issue as growth continued apace.

In our view, Oxfam must take its share of criticism for some of these failures. Oxfam ought to have provided closer support and monitoring to enable the organisation to set and meet realistic strategic objectives, and not to over-stretch itself. Over-stretching and poor follow-through had always been a characteristic of the movement to some degree. However, they became much more serious from the time that the work of BOJBP began to be recognised with various awards (see below), which probably contributed to the leaders becoming over-ambitious – not for themselves personally, but for the programme. There was also seemingly no good reason why a joint review of BOJBP's work, which was first mooted in the late 1980s, did not take place until 1993, by which time several of the issues had become pressing.

Recognition and awards

In spite of all the criticisms made by Jagdish Pradan in his report, he went on to mention the organisation's significant achievements in putting a stop to both contract lumbering and illegal tree felling, using the Gandhian techniques of *satyagraha* and *dharna*.[11] He also noted the enormous respect that BOJBP had earned from various departments of the government of Orissa, which 'has added strength to their work. Otherwise, an organisation comprising... part-time volunteers perhaps would not have been able to work so effectively covering such a large area.'

In the late 1980s the movement won a number of important environmental awards. From the government of Orissa it received the *Prakriti Mitra* (Friends of Nature Award), and the Indira Environment Award. From the government of India it received the *Burkya Mitra* (Friends of Trees Award), which was personally presented by the Prime Minister, Rajiv Gandhi. The most prestigious of all its awards came from the United Nations Environment Programme, which in 1989 acknowledged the quality of its work by presenting it with the Global 500 Award, 'in recognition of outstanding practical achievements in the protection and improvement of the environment'.[12]

The Mahasangha

Despite all these significant achievements, it was beginning to be apparent in the early 1990s to many in the movement and in Oxfam that BOJBP could not continue to service the needs of its own mother area and the ever-growing needs of sister organisations and the IVFPCs. In April 1992, therefore, a convention of sister organisations and IVFPCs was held, at which it was proposed to set up a super-federation, a *mahasangha*, embracing them all. The purpose of this umbrella organisation was to coordinate the activities of the member organisations and 'to advise the government for the proper protection, maintenance and utilisation of the forest'. It gave itself a name: *Nayagarh Jangala Surakhya Mahasangha* (Nayagarh Forest Protection Federation). The role, development, and effectiveness of the *Mahasangha* are considered in detail in the next chapter.

Conflicts and their resolution

With the continued expansion of their work, the chief functionaries and key volunteers of BOJBP found themselves increasingly drawn into disputes within and between villages, both in the mother area and in the wider community of sister organisations. These disputes frequently arose when one group of people (village or caste) was lax in its interpretation of the agreed rules and the other was strict, or more usually where some people in the community were finding it difficult to abide by the rules, because they were disadvantaged by them.

The Hazaris tell the story of a dispute in August 1985 between the villagers of the little settlement of Binjhagiri, on the other side of the hill from Kesharpur, and those of Nagamundali, just to the north. Three people from Nagamundali started to cut grass for fodder on the hillside above Binjhagiri village. Concerned that this would affect the stability of the soil on the hillside, the Binjhagiri people asked the cutters to stop. But they refused. Fearing a conflict with a village much bigger than theirs in an encounter that might have turned violent, they decided, against Jogi's advice, to take the case to court. However, while they were spending time in the lawyer's office and money on his fees, Jogi decided to appeal to the grass-cutters directly by prostrating himself. The move was successful: 'they became very repentant and their leader wept'. For the time being, at least, further conflict was avoided.

With reference to this and other incidents, the Hazaris observe that taking cases to court – or anywhere else outside the community – does not solve disputes: it only exacerbates them. They write:[13]

.... we have been highlighting the point that, come what may (even if the forest is destroyed), in no case should people go to the court or the police or the Forest Department for preventing the destruction of the forest ...
The intervention of an external agency will poison the atmosphere, put the village unity under great strain, drain the village of its resources [in lawyers' fees] and ultimately not only would the forest be lost but also the village would be lost. Hence the approach of the movement in crises has been one of motivation, persuasion, tolerance and understanding.

Tolerance and understanding, however, were extremely difficult to achieve where the interests of some groups (usually, but not always, the poorest) were ignored in the wider cause of environmental protection. The 1994 ODA Report cites the case of a group of villages within a sister organisation where there were 35–40 potters who were dependent on the nearby (now protected) forests for fuel for their kilns. From 1985, BOJBP advised the sister organisation to allow an area of forest specifically for the potters to use, but this advice was rejected. As a result, the potters were forced to purchase wood from the Forest Corporation depot and collect brushwood from a forest – someone else's forest – 10 km away. Even then they could not get enough fuel to produce pots at their former level of output.

Although in this particular case the disadvantaged group did apparently abide by the rules, it is not difficult to see that in other circumstances some people would be very tempted to continue breaking them. A recent report prepared for Oxfam[14] cites a case of the *dalit* village of Bebarthapalli near Nayagarh, where 75 per cent of the households are partially dependent on firewood sales. In spite of the pressure from the sister organisation and from neighbouring villages, wood-cutting has persisted here, simply because the community believe that they have no alternative livelihood resource.

Could conflicts have been avoided?

This raises an important question: how is it that a non-violent organisation such as BOJBP, founded on Gandhian principles, did not anticipate and prepare for conflicts? To understand this more fully, one needs to look more closely at the character and priorities of BOJBP. Throughout its history, the movement has always had a number of 'biases', which have unintentionally contributed to the creation of conflicts. The Oxfam report cited in the last paragraph suggests four biases.

1. Because BOJBP's outlook and foundation are Gandhian, it is not predisposed to a class-based analysis of development issues within which the

interests of the poor and the rich are differently analysed. As a result, the impact of forest protection on the very poorest and most vulnerable groups is generally not considered; or, if it is, it is not considered fully enough. Even in cases where it is taken into account, like that of the potters cited above, that consideration does not lead to adequate affirmative action on behalf of those groups or communities. Time and again the interventions of BOJBP people, of which several have been described in this chapter, have involved appealing to people's sense of and commitment to 'the greater good' as the means of resolving a conflict, rather than having in place and activating a conflict-resolution procedure. As seen in the case of the *dalit* village of Bebarthapalli, working for the greater good would have so seriously compromised their livelihoods that they were not prepared to go along with it.

2. BOJBP's focus of concern has always been, first and foremost, the physical environment, rather than the needs of any particular social group. Since its goal is environmental protection, it has had to seek the co-operation and involvement of all groups using the forests, regardless of how much or how little they used them and depended on them. As a result, protection of the forests, particularly in the short term, affected most adversely those who used them most, and they were always the poorest and most vulnerable.

3. Although the better-off[15] families were probably the least dependent on the regenerating forests, their co-operation and involvement was sought, not just because of the need to have everybody 'on board', but because of their status, which strengthened the hand of BOJBP when dealing with other villages and with external agencies, such as the Forest Department. Thus, the better-off have often held power out of proportion to the benefits which they have derived from forest protection, and conversely the poor have often held less power.

4. Finally, many village forest-protection initiatives have arisen when the forests have become so seriously degraded that the very poorest people have difficulty in meeting any of their needs from forest resources. When things have got so bad, both the poorest and BOJBP have seen forest protection as a 'win–win' situation, one in which everyone can gain, and very few, if any, lose. However, as we have seen with the potters and with the villagers of Bebarthapalli, unless special concessions are made, it is not necessarily win–win for absolutely everybody. Furthermore, as observed in (2) above,

until the forest has regenerated, it may not be a 'win' for the poorest for several years. In fact, in many instances the years of sacrifice could be endured only through 'raids' on other people's forests where protection had not yet started. And that, of course, could lead to further conflicts.

It would be wrong to deduce from all this that BOJBP's approach to environmental protection has inevitably produced conflicts incapable of peaceful resolution. While there is growing criticism of BOJBP's inability, even unwillingness, to help to resolve some of the thornier long-standing disputes, it is important to acknowledge that within and between many villages conflicts have been dealt with by the villagers themselves, sometimes with, but often without, the intervention of BOJBP or the *Mahasangha*.

Summary

Whatever difficulties began to emerge towards the end of BOJBP's first ten years of existence, there is no doubt that its successes were very considerable indeed. From the initial 22 villages with which it began in 1982, the movement's message of forest protection and regeneration had spread to well over 1000 villages by the early 1990s. Among these there was a close network of more than 300 villages, clustered into 18 sister organisations, plus a number of independent village forest-protection committees, all of which BOJBP supported and co-ordinated. Out of this in 1992 the federation, *Nayagarh Jangala Surakhya Mahasangha*, was created, initially to do campaigning work and later to co-ordinate the programme. What is most impressive is the level of work and commitment which the volunteers and staff of BOJBP were able to maintain, to keep the wider movement going for so long. That in the end it became difficult to sustain should perhaps not be surprising, given their modest resources. The next chapter will examine the current problems and the challenges faced by the organisation as it moves toward the end of its second decade.

Problems and challenges 6

'We do not want to enter into a partnership with the Orissa Forest Department or anyone else for that matter. The forest is ours and we have nurtured it for four decades, from the time when there were only some stumps left. Now the forest sustains all of us. Where were these help givers when there was no forest? We simply cannot let them enter our forest. We have protected our forest not only to sustain the present generation but also the succeeding ones.'
(quoted by Amit Mitra and Sanjoy Patnaik[1])

A new beginning

As we noted in the last chapter, in 1993 the ODA and Oxfam conducted a comprehensive joint evaluation of BOJBP, the first of its kind, which identified many issues that needed to be addressed and which contributed substantially to the restructuring of BOJBP. The evaluation team made the following recommendations.

- BOJBP should not extend its area of operation; rather it should consolidate its efforts in the current 324 villages.
- It should focus on strengthening the sister organisations and the *Nayagarh Jangala Surakhya Mahasangha* (the Nayagarh Forest Protection Federation, or Mahasangha).
- It should concentrate on supporting people's organisations to develop problem-solving and benefit-sharing systems.
- It should also focus on the involvement of the poorest sections of the communities in management and decision making on issues affecting them. If the livelihood of the poorest is adversely affected by any act of the project, then BOJBP should try to arrange alternative means of livelihood for them.
- It should aim for gender integration at all levels, including the involvement of women in decision making at the local level.
- It should strengthen its human resources, which would mean mobilising more unpaid volunteers, recruiting more paid staff, and training staff,

volunteers, and representatives of the sister organisations and the independent village forest-protection committees (IVFPCs).
- Through campaigning and lobbying, the *Mahasangha* should seek to influence government policy affecting forest-protection work.

Acting on these recommendations, BOJBP went ahead with strengthening the activities of the sister organisations and the *Mahasangha*. More staff were recruited, and more resources were mobilised from Oxfam for activities at both levels. Also serious attention was given to enhancing women's participation in environmental work. To meet these objectives, BOJBP drew up an ambitious list of activities for the period 1994 to 1998:

- Resolving conflicts, both within and between villages, arising mostly out of forest-protection activities and the use of forest resources, but arising sometimes from endemic factionalism.
- Campaigning on the environment and forest protection through *padayatras*, awareness meetings, camps, and the distribution of leaflets and other education materials.
- Running environmental education programmes in schools, to sensitise children on the need for environment protection through lessons, poster exhibitions, competitions, cultural activities, and campaigns.
- Holding teachers' workshops for an effective environmental programme in schools.
- Forming village women's groups for savings and credit, income generation, environmental protection, and tackling violence against women.
- Strengthening women's participation in forest and environmental protection through awareness camps, training programmes, workshops, exposure visits, demonstrations, plantations, and herbal gardens.
- Training youth volunteers in environmental protection.
- Protecting wildlife through workshops and awareness camps for hunters.
- Organising cultural activities in the villages for sensitising people to environmental problems and the need for protection.
- Organising workshops for artists and performers involved in cultural programmes, in order to develop environmental education packs, using cultural activities as a medium.
- Training in the management of tree nurseries; developing a permanent

nursery at Kesharpur; and promoting nurseries in schools.
- Celebrating *Banamahotsava* ('Forest Festival') and using it to promote plantation raising.
- Promoting the seed bank to meet people's needs in growing useful varieties of trees.
- Promoting forest-produce marketing initiatives through the *Mahasangha*.
- Campaigning for the protection of the Social Forestry plantations and for fair prices for their produce.
- Strengthening the *Mahasangha* and sister organisations through regular meetings and campaigning activities, such as *padayatras*, cultural programmes, training programmes, workshops, exposure visits, and information exchanges.
- Promoting environmental sanitation through awareness camps and the construction of safe latrines.
- Promoting good health through eye-operation camps, leprosy check-ups, and general health check-up camps.
- Holding workshops for pastoralists and other interest groups on the need for environmental protection, and what their roles should be in it.
- Organising a small programme of integrated tribal development in the area covered by BOJBP and the *Mahasangha*.

The Nayagarh Jangala Surakhya Mahasangha

Creating a new caste

We turn now to the formation of the *Mahasangha*. The concept of the *Mahasangha* emerged from a meeting of an innovative *Jatiana Sabha* ('caste group') held in Kesharpur in early 1992. At this meeting the notion of a 'forest caste group' – cutting across and transcending the existing caste structure – was discussed. It sprang from the idea, borrowed from west Orissa, that everyone involved in environmental protection and regeneration had a commonality of purpose and commitment which could be translated into a commonality of caste: 'What we are doing together unites us more than our castes divide us. We are therefore a new caste: a "forest caste".' Before the meeting, a leaflet entitled *Ye jatiana sabha kanhiki?* ('Why this caste-group meeting?') was widely circulated. The meeting received an overwhelming response from the people, and figuratively speaking laid the foundation stone of the *Mahasangha* as a popular organisation based upon a sense of unity and solidarity on forest-related issues.

In April 1992, a two-day conference was held at Muktapur, near Nayagarh, at which the idea of the *Mahasangha* was taken forward. BOJBP's work report for April 1991 to April 1992 reads: '.... BOJBP felt the necessity of a central committee (a Forest Protection *Mahasangha*) to be formed. The aims of the *Mahasangha* are to solve the day to day problems [of local communities] and to keep them active and to advise the Government for proper protection, maintenance, and utilisation of forest.'

At this meeting an *ad hoc* committee was formed, with Raj Kishore Subudhi from one of the sister organisations as its first President, and every sister organisation and IVFPC in the BOJBP network entitled to membership.

Clarifying the functions of the Mahasangha

Over the next three years, from 1992 to 1995, the *Mahasangha* with BOJBP attempted a number of things:

- It broadened the base of its membership, aiming to be as inclusive as possible by involving all forest-protection communities in Nayagarh District, plus any individuals who had a commitment to forests and environmental protection.
- It started to act as a bridge – a liaison organisation – between forest-protecting communities, giving them opportunities to raise issues, share their experiences, and learn from each other.
- It provided them with a platform on which to express their concerns about government forest policies and practices.
- It started public campaigning, though rather sporadically, on forest protection.
- It started work on the resolution of conflicts within and between forest-protecting communities.
- It began to co-ordinate the work of the nine strongest sister organisations, for each of which BOJBP, with Oxfam support, recruited and trained a Field Officer. While BOJBP continued to keep a close eye on the activities of these sister organisations, they were now 'autonomous', and reported to a Co-ordination Committee of the *Mahasangha*, which was set up in November 1995, rather than directly to BOJBP. In addition to its co-ordination function, this committee was involved in a process of 'mutual monitoring' with the sister organisations.

It must be stressed that, although the *Mahasangha* had its own objectives, it was still an arm of BOJBP. BOJBP saw it as its project or 'offspring' with, from October 1994, Jogi as President looking after it on behalf of BOJBP. In order to help the Executive Committee of BOJBP to understand the purpose of the *Mahasangha*, three workshops were held. At that stage there was no suggestion that the *Mahasangha* would become independent: no one ever imagined that it would act outside BOJBP's ultimate control. That is why, with separation now a reality, most of the chief functionaries cannot and will not accept the *Mahasangha*, and they blame Jogi for what has happened: an issue to which we return later in this chapter.

Sorting out structures and objectives

In 1996, at a further *Mahasangha* conference, a three-tier structure was adopted, consisting of a General Body, a Representative Council (*Pratinidhi Mandal*), and an Executive Committee. The General Body consisted of individual members of sister organisations and IVFPCs, and other concerned individuals, who each paid a Rs.1 membership fee. (Some 1580 people have now become members through this process.) The Representative Council consisted of five nominated representatives from each sister organisation, and two from each IVFPC, totalling 166. From this body 48 people were elected to the Executive Committee, responsible for the management of the *Mahasangha*. In order to streamline its functions, three sub-committees were formed. One was to provide legal advice; the second was to resolve conflicts emerging from community-based protection and management of forests in the District; and the third was to produce and circulate materials to sensitise forest-protecting communities and their networks. It was also decided to form other sub-committees if necessary, for example for fundraising or human-resource development. Oxfam undertook to facilitate training programmes for the Executive Committee.

Later in 1996, arising from further debates on the structure and functions of the Mahasangha, the leaders adopted the following resolutions and refinements:

- The Executive Committee would be constituted with a tenure of three years and have at least 15 members. The selection of members of the Executive Committee would be done through a democratic process. Marginalised groups, such as tribal peoples, Scheduled Castes, and women, would be properly represented on it. It was also agreed that individuals outside the sister organisations and IVFPCs could be nominated on a co-opted basis to the Executive Committee by the Representative Council.

- Through the conflict-resolution sub-committee, the *Mahasangha* would identify and respond to disputes as an important area of intervention, since unresolved conflicts are a severe threat to the continuing survival of the forests.[2]
- Campaigning against 'anti-people' policies, laws, and programmes in the forestry sector would form the core of the *Mahasangha's* work, and it would lobby the government to change policies and practices. In particular, campaigning against the harvesting of Social Forestry plantations would be boosted.
- Existing women's associations would be strengthened through awareness raising, education programmes, and organisational development.
- Efforts would be made to build the capacity of the key players in the *Mahasangha* through training programmes and staff development.
- Links would be established with organisations and networks working on the same issues in other parts of Orissa.

In 1996, after its redefinition and restructuring, the *Mahasangha* moved its office from Kesharpur to Nayagarh. It did this in order to be more functionally central within its operational area, and to be in closer touch with all the sister organisations and IVFPCs, and also with government agencies at the district level. From this location they became better placed not only to co-operate with existing member organisations, but also to form and develop relationships with new ones. Between 1995 and 1998, the staff of the *Mahasangha* formed a number of new sister organisations. Now there are 24 sister organisations and 23 IVFPCs, embracing some 400 forest-protecting villages. (The number of villages within the individual sister organisations ranges from 5 to 36.)

Some issues in which the Mahasangha became involved

As we noted in Chapter 2, such is the state of Orissa's forests that by and large most Protected Forests have been completely wiped out, leaving only Reserved Forests worth protecting, though many of these are also in a highly degraded state. This means that, while some villages in the *Mahasangha's* operational area are protecting *Khesra* forests (that is, Undemarcated Protected Forests), most are protecting Reserved Forests, within which according to the law people have few – if any – rights of usage ('usufruct rights'). But to local communities these official designations are unimportant. What counts for them is simply the presence of a forest on which they have depended and which they have protected – irrespective of legality. To assist sister organisations and IVFPCs, the *Mahasangha* decided that it was important to make forest-protecting communities fully aware of their rights over their local forests. They also campaigned on the issues of Social Forestry and the advantages and disadvantages of Joint Forest Management, both of which we consider later.

In addition to all this, the *Mahasangha* took a stand against the Forest Department when it launched a drive to get people from forest-protecting villages to undertake thinning and cleaning of the forests, some of which was taken up as part of a cash-for-work drought-relief programme. The *Mahasangha's* position on this is that, since these forests are under the *de facto* control of local communities, working under community-based forest-management arrangements, then the Forest Department has no part to play in them whatsoever – doing thinning and cleaning, or anything else for that matter. Furthermore, as far as local communities are concerned, by and large, thinning and cleaning is not appropriate to the way in which they generally manage their forests.

The *Mahasangha* has taken steps too to stall the harvesting of bamboo by the State Forest Development Corporation in those forests that are under effective community control. Bamboo, a nationalised product, is found in almost all types of forest in Orissa. The Orissa Forest Development Corporation is a State undertaking, operating under the Forest Department. It is responsible for harvesting timber, fuelwood, bamboo, and other non-timber forest produce. It has a monopoly over the harvesting of bamboo from all types of forest, which it sells to the paper and pulp industries. In Nayagarh District some forest-protecting villages have for years been harvesting bamboo to meet their local needs. Seeing no reason why this should change, the *Mahasangha* has sensitised villagers on the issue. As a result, when the Forest Corporation staff come to harvest bamboo, their access is barred by people claiming, 'We have been protecting this bamboo. Why should you now come and harvest it to cater to the needs and demands of the paper mills?'

The drive for greater democracy and effectiveness

Towards the end of 1997, after a further process of informal self-assessment, the *Mahasangha* decided that there had to be greater internal democratisation in order to make it a genuine people's organisation. They felt that it still had not become democratic, despite all their attempts to broaden its base. Discussions were held with the sister organisations and some of the IVFPCs on a number of issues. Subjects included the purpose of the *Mahasangha* and how it could be made into a real people's front; resource mobilisation;

self-reliance and self-sustainability; and devising rules for usufruct sharing. The majority of the sister organisations and the IVFPCs agreed that there needed to be a more effective democracy than had been achieved to date. Further discussions were continued at the village level, and so far more than 150 such meetings have been conducted, to discuss people's aspirations for the *Mahasangha* and solicit their ideas for making it effective and sustainable. Another 100 or so meetings are planned.

In January 1999 the members of the *Mahasangha* held a convention at Kalyanpur and developed a plan for further restructuring. It was decided that four tiers were needed; at the levels of village, sister organisation, zone (a cluster of sister organisations), and *Mahasangha*. From each village, five members, including two women, will be elected to the sister organisation; three people, including one woman, will be elected to the zonal federation; and one man and one woman will go to a Representative Council at the *Mahasangha* level, from which an Executive Committee would be chosen.

The convention also drafted plans for raising resources for the *Mahasangha*. It was proposed that two adult members from each family in each forest-protecting village should become 'members' of the *Mahasangha* and each should pay Rs.1 as a membership fee per quarter (Rs.8 per family per annum). The money will be paid initially to the village councils (and IVFPCs), and portions of it will then go up through the structure to the sister organisations, zonal federations, and the *Mahasangha*. The exact size of these portions will be decided when all villages have been consulted.

The *Mahasangha* is presently trying to mobilise new leaders and volunteers, who can take forward the process of network-building. Simultaneously efforts are being made to mobilise women and children in the process of networking.

The *Mahasangha* has recently become a pioneering player in the development of the *Odisha Jangal Manch*, a State-level forum of forest-protecting organisations. The Secretary of the *Mahasangha* is the forum's convenor. Before the 1999 national elections, the forum circulated an agenda for forest protection, to be included in the manifestos of the political parties, thus using Indian democracy to advocate for change.

Promoting gender equity in the environmental movement

Group formation

BOJBP's activities for women's development did not seriously start until 1989, and were brought into sharper focus in 1995, following the recommendations of the 1993 ODA/Oxfam joint evaluation. As we noted earlier, the evaluation recommended that gender integration needed to happen at all levels within the organisation.

During the period August 1994 to June 1995, BOJBP organised approximately 70 sensitisation camps, attended by more than 1000 women in the operational areas of BOJBP and three sister organisations, *Basundhara Bandhu*, *Ratnamala Jungal Surakhya Committee*, and *Sabuja Jeevan*. Workshops were conducted at selected places to plan strategies for developing women's groups at the village and supra-village levels. To this end they worked out another four-tier structure for federating women within Nayagarh District:

1. at the village level, there would be a women's front, called *Paribesia Mahila Surakhya Vahini* (PMSV: 'Women's Brigade for Environment Protection');
2. at the village-cluster level, an association of PMSVs would form an *Upamandal* (sub-council);
3. at the level of the sister organisations, three or four *Upamandals* would form a *Mandal* (Council);
4. at the district level, once Mandals had been formed, there would be a *Mahasangha* or *Mahamandal*.

Thrift and credit schemes

A Women's Co-ordinator joined BOJBP in November 1994 to help to scale up the activities of the Women's Development Programme. First, BOJBP urged all sister organisations to form PMSVs, and to improve their own structures to facilitate women's involvement in management and environmental protection. Second, working with their own villages and with those of three sister organisations, they embarked on a programme of thrift and credit promotion within the PMSVs. They saw this as a strategic way of bringing women together in a shared activity, and of uniting and organising them for economically productive activities. Oxfam played a strongly supportive role in this, developing the capacity of the staff and of the credit-group members. This they did through training, orientation, and facilitating exposure visits to other thrift and credit organisations in Orissa and Andhra Pradesh.

By February 1999 there were about 80 PMSVs, with a membership of 1945 and a combined capital of Rs.435,000 (about £6,500). These PMSVs have developed nine *Upamandals*, and in the operational area of *Sabuja Jeevan* one *Mandal* has been constituted. Now thrift and credit activities have been rapidly taken up in three more sister organisations. The whole process has facilitated at least 70 women to start small enterprise activities, including the processing of paddy, growing vegetables, selling vegetables and other foods, processing food, and fruit-squash making. Women also borrow money to meet exigencies related to ill health, marriage, and their children's education. Initially women were hesitant to join the thrift and credit groups, because of a lack of trust and confidence. Now there is a growing interest and even competition among them. In many villages new groups have been formed, and some have already saved more than the longer-established groups. While there are still conceptual and organisational issues to be resolved, the programme has certainly provided women with opportunities to develop a platform for their own solidarity and development.

The growth of confidence

Besides thrift and credit activities, the PMSVs have undertaken a variety of environmental protection work and village-development activities. One has been environmental sanitation, which includes cleaning village roads and the areas around open wells and tube wells. Another has been personal hygiene and preventative health care.

Women in PMSVs have also adopted various forms of protest, including hunger strikes to resolve cases of domestic violence. As a result, cases of violence against women have been substantially reduced. Now, if a man tries to prevent his wife, sister, or daughter from attending a PMSV meeting or taking part in a PMSV programme, the members will usually 'counsel' him. If he remains obstinate, the women will organise a protest in the village –'naming and shaming' the offender – to draw everyone's attention to the problem.

There is a growing number of cases in which women have been involved in communal development activities, particularly in lobbying for government assistance and funds, and in stirring village councils to sort out village problems. The story of the women of Talapatana illustrates what women's power and drive can accomplish. It was told us by Alaka Nanda (BOJBP's Women's Co-ordinator):

> In Talapatana, one of the hamlets which makes up the village of
> Chadhayapalli, the women have a very interesting story to tell ... This hamlet did
> not have a road to connect it with the main village ... The villagers were frustrated
> in their attempts to woo the block and *panchayat* officials to construct the road.
> The problem was non-availability of government land on which the road could
> be constructed. The women raised this issue in the *Upamandal* meeting.
> They tried to motivate the people owning land [along the route of the proposed
> road] to contribute it for larger village development. But the landowners asked
> for compensation. Failing to motivate them, the women decided to go on

hunger strike. On 26 May 1997, 42 women from five villages went for a prostration in Chadhayapalli from 4 am. In the morning, at the sight of this the villagers became horrified. Old people in the village could not control their tears. The village road was immediately covered with an awning to save the women from the heat of the sun. After four hours, the landowners agreed to part with their land, and Ratnamala Jungal Surakhya Committee promised to pay minimum compensation to them. By noon the land for the road was demarcated and the demonstration was called off. The work, which couldn't be done for the last 15 years, was made possible by the women, once they got themselves organised. **'**

By itself this single episode does not mean that women have suddenly become organised and empowered, but, along with stories of other protests, it does reflect both the potential and an emerging trend of women's involvement in development activities in the area.

Women in forest protection

It is also true that women's awareness of environment protection and regeneration has increased considerably, though they still have a long way to go in terms of organising for collective action. There is a growing interest among them in developing plantations of useful trees, both around their houses and on the village commons.[3] And although women are still rarely part of forest-protection committees at the village level, in many places they have actively contributed to forest-protection work. There are even examples of women taking part in *thengapalli*, and in one remarkable case doing it on their own. This is what Alaka Nanda and Sitansu Sekhar Mohapatra (Field Organiser from the *Mahasangha*) told us:

> **'** Hariharpur village, near Sarankul, is the centre of *Dharitri Bhandara Surakhya Parishad*, a sister organisation of BOJBP. During the period December 1996 to January 1997, because of a conflict within the village, people neglected their protection of the local forest, which provided an opportunity for some to start chopping down trees. Failing to stop the destruction, the women of PMSV in Hariharpur, with strength of 30 active members, started a *thengapalli* to protect the forest. By this they were able to control the illegal felling for a period of 22 days. Then representatives of *Dharitri Bhandara Surakhya Parishad*, BOJBP, and the *Mahasangha* went to the village and motivated all the villagers to protect their forest. The conflict was resolved, and the protection by the village committee resumed. **'**

Some women leaders are showing increasing interest in attending meetings organised by BOJBP and the *Mahasangha*. In the last two annual *Mahasangha* conventions, women have attended in their hundreds, and some of them have also addressed the public gathering. In one sister organisation, *Sabuja Jeevan*, mentioned earlier, women from ten PMSVs have formed three *Upamandals* and one *Mandal*. The leaders of the *Mandal* are active in attending the meetings of BOJBP and the *Mahasangha*. They are also involved in the programmes of *Sabuja Jeevan*, the sister organisation itself. Facing a growing demand from the *Mandal* to take women members into *Sabuja Jeevan*, they decided to nominate women leaders from three villages on to the Executive Body. But meetings are usually conducted in the evening, which makes it difficult for women to participate. So a decision has been made to organise the Executive Body meetings in three villages on a rotational basis, so that women leaders from at least one village can attend. The leaders of the *Upamandals* in the Chadhayapalli area have made similar demands about becoming members of their sister organisation, the *Ratnamala Jungal Surakhya Committee*, which as a result is now planning to take more women as members. Their primary demand is that they should jointly manage the forest with the Committee.

The women of the small village Binjhagiri, in the BOJBP 'mother area', about which we have told several stories, have shown extraordinary concern for forest protection. In 1997, the women there raised a plantation in the village forest. They fenced the area, watered it every day, and kept a close eye on it, pleased to have forest resources within easy reach. One day they decided that the patch needed 'cleaning' and they asked the men of the village to help, but in vain. So the women started to do it themselves, and cleaned up fully one acre. This shamed the men into action, and they cleaned up rest of the area. Speaking of the way in which women's attitudes have changed, Kaushalya Pradhan, Kumudini Maharana, Aparna Swain, and Kanaka Barik – all Binjhagiri women – told us:

> ❛ Things have changed. Women have come forward to participate in public meetings and to interact with outsiders. They can now go to far-off places to attend programmes of BOJBP, leaving behind their household work. No woman can be harassed in our village. Through our challenge we can protect her. ❜

During 1994–95, women attended several programmes organised by BOJBP, and many came forward to participate in a number of strenuous demonstrations and *satyagrahas*. One of these took place in mid-1994 in the faction-ridden village of Badagarada, just down the road from Kesharpur.

Failing to resolve the conflicts, BOJBP decided to hold a mass prostration. All villages in the core area of BOJBP and in the sister organisations, plus the IVFPCs, were contacted, and a day was fixed for holding a mass *satyagraha*. More than 200 people, including 10 to 15 women, came and slept on the village road for several hours, until the villagers of Badagarada decided to resolve their disputes and restart forest protection. This was a 'first': never before had women from villages outside the immediate area taken part in such a protest. It set an important precedent.

Obstacles to integrating women into the movement

Although all of these accounts show that there have been significant successes in the involvement of women in forest regeneration and protection, still it has to be said that the record is mixed. It is true that in many villages PMSVs are functioning well, and thrift and credit activities have made a good start, with the support of both women and men. But, while women's environmental awareness has been raised considerably, in general they have not been involved in action on a regular and sustained basis. This continues to be an issue, therefore, within both BOJBP and many sister organisations, which can be accused – still – of not taking gender equity seriously enough.

Although a Women's Co-ordinator and women staff have been recruited to scale up activities, resource allocations for their work have not been adequately planned and utilised. Furthermore, no serious efforts have been made in participatory planning with women's groups. Nor have they been given a chance to articulate new ideas for intervention, or even blend existing women's development work with the environment-protection movement. Generally women are still seen as a separate social group, fighting in isolation for peripheral changes in their status. With some notable exceptions, women have been integrated into the decision making at neither the village level nor the level of sister organisations and BOJBP – that is even after five years of 'organised interventions'. In the 101-strong General Body of BOJBP there are still only eight women, and in the Executive Body there are just two. Furthermore none of them is active in the organisation, and no efforts have been made to make them so. Even in Kesharpur women are not effectively organised, and they have not regularly participated in any of the activities of BOJBP, apart from those relating to savings and credit. While it is possible to point to sister organisations, such as *Sabuja Jeevan*, where change has come, such examples are still too few in number.

The major problem for the staff is the lack of proper guidance and direction for the development of an effective strategy for women's empowerment. Women are simply not involved in developing plans. They are always asked to carry out instructions, which may sometimes contradict the decisions of women's groups. Furthermore, little flexibility is allowed to the staff involved in the Women's Development Programme. Integration of thrift and credit work into the mainstream environmental movement for which the programme was designed is still lacking. Rather the organisation is making headway for establishing thrift and credit as a separate programme, with plans to set up a women's bank.

The recent conflict between BOJBP and the *Mahasangha* has aggravated the situation. Now BOJBP intend to separate the Women's Development Programme, which they still feel is 'theirs', from the sister organisations. Participation of women's groups in the *Mahasangha's* activities is resented by BOJBP. In the *Sabuja Jeevan* area, this resentment has become explicit. There the staff of BOJBP have persuaded the leaders of the *Mandal* not to attend the meetings of *Sabuja Jeevan* nor those of the *Mahasangha*. Instead of collaboration, it is intending to develop parallel institutions, which makes a nonsense of their aspirations for integration.

The problem of sustainability

As we have seen, when the movement started, BOJBP's focus was on campaigning, spreading the message of environmental protection. The aim of each of the core villages was to see 'their' hill re-clothed with forest and their farmlands restored to full fertility and productivity. No thought was given to what should be done with the forests once they were restored: whether they should simply continue to be protected, or whether they should or could be harvested on a manageable basis. At the most, it was assumed that, once they had reached this state, people would fulfil their subsistence needs on an *ad hoc* basis, taking no more than they needed. BOJBP failed, therefore, to develop any kind of comprehensive usufruct system.

The present challenge for BOJBP – and for the sister organisations, IVFPCs, and the *Mahasangha*, too – is to introduce an equitable benefit-sharing mechanism to cover all forest produce. At the same time, for BOJBP, there are three other challenges that compound the problem. There is growing factionalism and conflict within and between the villages; a certain break-down of people's trust in BOJBP; and rising disaffection in the younger generation, which has little stake in the movement or interest in forest protection. In spite of the fact that some villages in the mother area have introduced their own rule systems for forest usage, 'illegal' felling is on the increase. Particularly intractable has been the dispute between Sanagarada and Badagarada, which has led to quite serious forest destruction on the slopes behind them. As a result, the core villages have considered demarcating 'their' patch of forest to exclude outsiders. (Kesharpur has already done this: see Figure 11.) This has generated animosity with the other 13 villages, which have for years restrained themselves from taking from the forest and now think that they should derive some benefits from it. The situation is made all the more precarious because the type of indigenous forest which grows here, that is 'Dry Mixed Deciduous', is relatively slow growing. If a serious, all-out assault were to take place, the timber would all be gone within two years.

The good news is that some sister organisations, including *Sulia Paribesh Parishad Anchalik*, *Sulia Bana Surakhya Committee*, and *Ratnamala Jungal Surakhya Committee*, have developed usufruct-sharing mechanisms which are working. These organisations have introduced commutation fees ('royalties') for various types of forest produce, but only for *bona fide* use. Every year the *Ratnamala Committee*, based at Chadhayapalli, harvests bamboo from both Reserved and Protected Forests, and sells it to member villagers and to outsiders at the same prices. This is happening in spite of the fact that bamboo is a nationalised product, and harvesting it is illegal. The Forest Department knows what is going on, but so far has turned a blind eye.

Figure 11: A recent and serious attack by some villages on some parts of the regenerated forest on Binjhagiri has led the people of Kesharpur to put down boundary markers, painted white, in order to delineate and protect 'their' part of the hill This unfortunate development has arisen in part because a failure to institute a sustainable system of benefit sharing among all 22 villages.

In these and other places in the area, forest-protecting communities have developed informal rule systems to use the forest according to their needs and the forests' capacity. This is a good start, but nobody has seriously tried to define the needs of the people and their demands on the forest, nor the needs of different groups within those populations. Neither has there been any attempt to assess the potential of the forests to meet those demands, so that sustainable management systems can be devised. Recently, the *Mahasangha* has started to develop strategies for the management of NTFP in different parts of the District, but as yet it has made little progress.

The response of BOJBP and the *Mahasangha* to the Social Forestry Project

In the mid-1980s, BOJBP was actively involved in the Forest Department's Social Forestry Programme, supporting forest officials in planting village wood-lots with community participation. Working with the Department, it organised a big campaign to establish these plantations, but they failed to collaborate to devise any long-term management plans. Planted with fast-growing exotic species, by 1995 many of these wood-lots were mature enough to be harvested. But still it came as a shock to members of the *Mahasangha* and BOJBP when some villages did actually harvest them, readily selling the wood to agents and contractors of paper mills and other industrial establishments. There are reports that some agents and Forest Department staff spread rumours that the timber would be useless if it was not harvested before it was ten years old. To 'oil the wheels' too, the Forest Department lifted the ban on the felling of trees and relaxed the rule that required timber and transit permits for certain species (normally needed to shift timber from the forests). These species were of course the ones grown in the Social Forestry wood-lots.

It is also said that various methods were adopted to 'persuade' village leaders to sell timber at throw-away prices. In one small village, Dhadhera in Nayagarh District, for example, a contractor successfully persuaded the villagers to part with all the trees in a 15-ha wood-lot for a mere Rs.12,000, the deal being finalised with an advance of just Rs.100 to the village committee. When the leaders of the *Mahasangha* heard about it, they rushed to the village and exposed the contractor's game. Around 42,000 trees had been planted in the wood-lot in 1984–85. Assuming a survival rate of 50 per cent, there might have been 21,000 trees left. The village had just agreed to sell them for the ridiculously low sum of Rs.0.60 per tree, when a single ten-year-old eucalyptus or acacia tree would be worth a minimum of Rs.20. Thus the village had just lost at least Rs.408,000 (well over £6,000).

Community forest management

Apart from crooked dealing, what this story illustrates is the subversion of the original objectives of the Social Forestry Programme. As we noted in Chapter 2, those objectives were to take the pressure off the forests by providing people with locally available firewood, small timber, and fodder, not to feed the gargantuan appetites of industry. This particular story, however, ends happily. Once the people of Dhadhera realised that they had been duped, they returned what the contractor had paid them, and stopped him from further felling.

'Beware the wood-eating tiger'

In order to prevent similar acts of deception being perpetrated elsewhere in Nayagarh and Khurda Districts, the *Mahasangha* and BOJBP launched an extensive campaign, using leaflets, posters, and booklets, and organising meetings and workshops in those villages with Social Forestry plantations. One poster, proclaiming '*Beware, the wood-eating tiger has come*', was widely effective and became very popular (Figure 12). This initiative, called the Forest Produce Marketing Initiative (FPMI), was aimed at protecting hundreds of acres of plantations from the ravenous appetites of scores of paper mills.

Figure 12: 'Beware: the wood-eating tiger has come' – or 'Don't be seduced by logging companies offering money for social forestry plots'. Social Forestry, a government scheme to promote wood-lots of harvestable trees, was designed to take the pressure off existing forests by providing accessible sources of domestic fuelwood and small timber. In reality it led to the mass planting of exotic species, which became the target of the loggers' axes. The BOJBP campaign against the enticements of the loggers was enormously successful.

In all, 117 villages were involved in the campaign, with representatives from each village constituting a Representative Council.

Not only did the FPMI campaign save huge areas of plantations, but it also bargained for a fair price for the timber and poles that were harvested. Village communities were trained in the management of the plantations, which involved rotational harvesting, replanting, and the equitable distribution and management of funds created by selling the surpluses from the wood-lots. A small booklet was circulated on 'how to raise seedlings'. Some 20 or so villages whose wood-lots had already been devastated by contractors were contacted, to motivate them to replant their wood-lots and to make best use of the money raised by the selling of the produce. BOJBP and the *Mahasangha* suggested that if a village did decide to harvest its plantation, which it was perfectly entitled to, it should keep back 25 per cent of the produce to be distributed equally among the villagers for their domestic use. They also suggested that 25 per cent of the money received from the sale of the timber should be spent on replanting; that 25 per cent should be kept in the village fund; and that the remaining money should be spent on village development initiatives, as decided collectively by the villagers.

There is no doubt that the FPMI has been immensely successful, in that it has enabled villages to stand up to the combined forces of industry and the Forest Department, and enhance their bargaining power for a fair and remunerative price for forest produce from the village commons. But FPMI has also added fuel to the on-going conflict between the Forest Department on the one hand and BOJBP and the *Mahasangha* on the other, a conflict which has been made worse by the issue of Joint Forest Management, to which we turn next.

The fight against Joint Forest Management

When in 1988 the State government announced its intention to involve village communities in the protection of Reserved Forests, there was no overt response by BOJBP. It was not until 1993, when the Joint Forest Management (JFM) Resolution was passed, that BOJBP began to appreciate the importance of the move. But it still took them some time – until 1995 – to gear up to campaign on the issue. Their first public action was to circulate leaflets explaining the problems of JFM, as they saw them. They also launched a signature campaign, demanding changes in the Government's JFM resolution, and in particular greater rights for communities; the withdrawal of a forester as secretary on each village JFM committee; and a 50:50 sharing of any harvest between the community and the Forest Department.

Whenever meetings were organised, BOJBP and the *Mahasangha* vehemently criticised JFM, and used the opportunity to warn people of the dangers of allowing forest officials into the domain of community forest management. Their basic argument was that for years people had been satisfactorily protecting forests through a community-based approach. Why, therefore, should there be a need for another approach, especially one that diminished their hard-won rights (even if those rights were still only *de facto*)? Basically they saw no need to replace their own system, in which they had control and authority, with one in which they would be unequal partners with the Forest Department, whose motives, in their view, were highly suspect.

During 1996 and 1997 the campaign against JFM was strengthened. Members and staff of the *Mahasangha* travelled extensively in the areas where the Forest Department was trying to implement it, explaining that the intention of the Forest Department was to gain gradual control over forest resources that had been regenerated and protected by village communities. They maintained that when a forester became the secretary of a forest-protection committee, the autonomy of the village in managing the forest and forest produce would disappear. They pointed out that even to take a single piece of fuelwood from the forest, villagers would technically need to ask permission of the Forest Department.

In spite of all this, following the passage of the JFM Resolution, some villages did come forward to participate, enticed by the promise of an income from the forest and *de jure* land rights over it. In response the *Mahasangha* felt that they had to redouble their efforts in campaigning against what they saw as false promises by Forest Department officials.

A break in the ranks

Things came to a head in 1996, when the *Anchalik Sulia Bana Surakhya Committee*, a sister organisation based at Machhipada, entered a JFM agreement with the Forest Department. This organisation consisted of 36 villages and about 2700 families. Collectively they had protected the forest on Sulia Mountain for several years, and harvested their daily needs from it. When BOJBP and the *Mahasangha* came to hear of the agreement, they were absolutely horrified, seeing it as a terrible threat to the community-based forest management that had been so painstakingly developed over the years. For if this sister organisation, which was one of the strongest in the network, could join hands with the Forest Department, then what, they asked, might happen with the rest? They arranged several meetings to discuss the issue, but people almost came to blows, and the sister organisation would not back down.

Then, in the summer of 1997, the sister organisation found itself with a price to pay for its collaboration. The members wanted to harvest bamboo from the forest on a regular basis, for which they had to get Forest Department permission. But under the new JFM arrangement, according to the Forest Department, only the 551 families of Machhipada village were members, and none of the families in the other villages. So when the committee approached the Forest Department on behalf of all the villages, the officials said that if the other villages wanted bamboo, they would have to buy it from a fund given by the Forest Department to the committee for the purpose: they could not themselves harvest it. The committee responded, arguing strongly that they had entered the agreement on behalf of all 36 villages, not just Machhipada. After repeated visits to the Forest Department, the latter finally agreed to allow all villages the right to harvest bamboo; but it had been a long and bitter fight.

Perhaps not surprisingly, the *Anchalik Sulia Bana Surakhya Committee* are now firmly against JFM and would like to have the arrangement cancelled. However, although technically they are still bound by it, throughout the whole of Nayagarh District the implementation of JFM is at a stalemate, through non-co-operation. To date only four villages, of which Machhipada is one, have signed up to it, and in none of the four cases is it working.

This collective experience has only strengthened the warning message of BOJBP and the *Mahasangha* concerning the pitfalls in store for villages engaged in community-based forest protection if they entered into a JFM agreement: 'Do not be lured by false promises of benefits. If you are, you will find that your rights will diminish and your control over forest resources will disappear.' This message has been reinforced in numerous articles in *Sabujima* ('Greenery' in Oriya), the newsletter of the *Mahasangha*. In addition, members of the Mahasangha have continued to attend State-level meetings, at which they have raised questions and objections to the whole principle of JFM.

The reaction of the Forest Department

So effective has the campaign been that JFM has been virtually killed stone dead in Nayagarh District. Wherever forest officials have attempted to initiate it, the local people have opposed them. This has obviously further inflamed feelings between BOJBP and the *Mahasangha* on the one hand and the Forest Department on the other. In an interview with us, Subash Chanda Mishra, a Section Forest Officer, Odgaon, said:

> BOJBP is opposing JFM. It is instigating people against the Forest Department. It has really become a difficult job for us to implement JFM.

BOJBP has formed a *Mahasangha*, which is creating problems ... They misguide people. But they can't do anything. As the conflicts in forest-protecting villages are increasing and they [BOJBP and the *Mahasangha*] are not capable of handling them, they have to come to us.[4] JFM is the only panacea. It permits the communities to legally use the forest produce. **'**

Another Section Forest Officer, in Pancharida, Purna Chandra Sahoo, told us: *'The anti-government stand taken by BOJBP is not good. It may drag the organisation into trouble. It is beyond the capacity of BOJBP to manage the forest successfully.'* In response to such comments, we were told by several people associated with the *Mahasangha* (Jogi, Laxmidhar Balia, and Hadubandhu Senapati):

> **'** We can't sell the birthright of our children for the interest of others. We don't want to get into any share-cropping business with the Forest Department. There can't be any equal partnership between the landlord and the tenants. If we spare an inch of forest for JFM, then that is gone forever. We can't revive the forest again. **'**

These are sentiments powerfully echoed in the quotation at the head of this chapter.

Internal challenges

There can be no doubt about the astonishing achievements of BOJBP and, in the few years of its existence, of its 'offspring', the *Mahasangha*. That they have become a powerful force for environmental awareness-raising and forest protection is recognised not only locally but nationally and internationally. However, it is also important to acknowledge that the movement now faces many internal challenges. In our discussions with the key functionaries, with the staff, with members, and with ordinary people, many of these challenges were revealed. We now look at them in some detail.

Voluntary labour vs. paid labour

First there is the issue of labour. One of the greatest qualities of BOJBP, which has always set it apart from most other organisations of its kind, is its spirit of and commitment to voluntarism. Unlike most voluntary organisations and NGOs, the volunteers really do manage it. The key functionaries are teachers and farmers, who take no money for their services. Although there are paid employees, the organisation's activities are planned and implemented by the volunteers with the assistance of the staff. The first paid employee did not join the organisation until July 1984; up to 1988–89, when two or three more joined, he was the only paid person supporting the volunteers in the movement. Then in 1992–93 the number of staff was increased to 11 as part of a three-year programme of scaling up, funded by Oxfam. Before that time, the paid staff could not really be considered as normal employees. For the most part they were operating as volunteers with a basic subsistence allowance. They worked no set hours, nor did they have job descriptions. Their work ranged from cooking and cleaning, to launching campaigns in the villages, to motivating people to undertake environmental protection. Since much of the organisation's work took place 'after hours', when the volunteer leaders were free, many of the paid employees worked all day and all evening, and sometimes, as when campaigning is at its peak, they work right through the night, too. Until 1992, the common perception was that, if you were with BOJBP, then you had no life beyond trees, forests, and the environment.

After 1994, BOJBP witnessed a gradual decline in the exemplary quality of voluntarism of the early years. This came about with a change in the volunteer leadership, which brought with it three things: an increasing dependence on paid staff; a reduction in the role of the volunteers; and a changed relationship – almost a role reversal – between the two. As the organisation scaled up, this was probably inevitable, but it was not foreseen and planned for. If it had been, the needs of staff and volunteers for training would have been identified, and management and support structures would have been properly worked out. Because this did not happen, many members of staff were ill-equipped to carry out the work. They lacked understanding and flexibility, and in the absence of proper leadership several did not have the right attitude to work. In some cases, when a staff member was attached to a sister organisation, he used his position to assert dominance over the volunteers. In some cases, too, staff were guilty of manipulation to boost their own power and influence. As a result, the volunteer leaders of some sister organisations became over-dependent on their staff, and there was a marked decline in their own commitment and involvement. Among some of these leaders developed the attitude: 'Since the staff are getting paid for the work, let them do it.' This did not happen everywhere, but there was certainly a shift in that direction. To some extent the same thing happened in the *Mahasangha*, too.

Perhaps all this was inevitable, given the volume and complexity of the work, and the full-time commitment which was required to do it. But it need not have happened to the extent that it did. The problem was essentially one of an obsolete management structure, in both BOJBP and the sister organisations, which was simply not geared to managing staff and volunteers in an inclusive way. They found that a strong belief in voluntarism does not in itself lead to the effective management of volunteers and staff. Managing human resources – particularly where there is a mixture of staff and volunteers – is a difficult task, which the organisation could not achieve without help.

It is no exaggeration to say that within both BOJBP and the *Mahasangha* the future of voluntarism is now at stake. Only those leaders who were in the movement at the beginning are continuing to show a high level of commitment. In the *Mahasangha* a new team of volunteers has emerged, but only time will tell if they can sustain the true spirit of voluntarism.

The loss of transparency and dynamism

Second there is the issue of openness, which people feel the organisation has lost in recent years. Many have referred to this problem, saying that BOJBP is now '*a closed organisation*'. One added, '*People are in the dark regarding the activities of the organisation, because of poor interaction.*' It is said that even the people of Kesharpur, who initiated the movement, are now not fully aware of what is going on.

Coupled with this there is an apparent loss of dynamism and creativity in the organisation. The General Body has become static, and some people feel that it consists of too many people who are no longer active or representative of the communities from which they come. The Executive Committee has been through a bad patch, with few meetings, little activity, and negligible changes in its membership. While there have been improvements recently – the Committee is now meeting every month – it still does not compare well with the times when, according to Susanta Jena and Priya Nilimani, who had been in the organisation from its early days, it met '*every day late into the evening to assess the work of the day and to plan for the next day*'.

At the same time, the participation of young people in the movement has declined, which has contributed to the loss of dynamism. In the early days there were many young people, but there seems to be a certain reluctance to involve them now, possibly, according to one of them, because of a fear that '*these youths might politicise the activities of the organisation*'.

A crisis of leadership

There has also been a crisis of leadership in the organisation, caused partly by the departure of Jogi. The origins of this date back to February 1996, when the *Mahasangha* moved its office from Kesharpur to Nayagarh and started to pull away from BOJBP. Until that time, Jogi, who had become President of the *Mahasangha* in 1994, was able to keep an active guiding hand on BOJBP. But when the *Mahasangha* moved, he would go every day after school to

Nayagarh to organise its activities. In so doing, he gradually moved away from the day-to-day activities of BOJBP. It is also possible that, finding that things in BOJBP were not moving in the right direction, he tried to develop the *Mahasangha* as an effective alternative organisation to promote community forest management throughout the District. Whether this is true or not, his departure left a huge gap, because he had played such a central role. '*There is only one Jogi*,' said one person to us. '*No efforts have been made to create other Jogis in the organisation.*'

The impact of these issues on the programme

A decline in volunteer commitment and the spirit of voluntarism; difficulties between staff and volunteers; lack of effective management; a certain loss of transparency and dynamism; the inadequate functioning of both the General Body and the Executive Committee; a loss of leadership and sense of direction; and the decline in the involvement of young people – all this has led to a 'drift' in the movement, which many acknowledge. Several who spoke to us feel that the objectives, ideology, spirit, and principles of the organisation have all been diluted.

That this has affected the programme cannot be doubted. Several told us that they thought the programme had lost its way and had shifted its central focus from environmental protection to other activities. They pointed to the establishment of a women's bank, the commencement of a tribal development programme, and the development of 'service delivery' activities, such as income-generation projects, sanitation, water harvesting and irrigation, and youth employment schemes. Others voiced concerns about the old problem, to which we have referred earlier, of undertaking too many activities that cannot be properly co-ordinated and sustained. One person expressed the view:

> ❮ There is no proper planning of programmes and activities in consultation with the staff and target communities. The staff are there only to carry out the instructions of the chief functionaries. There is no co-ordination among the staff to implement the programmes effectively. There is also a lack of proper communication between the leaders of the organisation and the staff. ❯

Finally, while the staff and volunteers are crucial for the execution of the programmes, no serious efforts have been made to develop their knowledge and skills. The staff have received inadequate support and they have not been given the flexibility that they need. Furthermore, right across the board, activities have not been properly monitored.

Clearly Oxfam must shoulder some responsibility for many of these internal problems. Until recently Oxfam has played a major role in helping BOJBP to set objectives, develop properly costed plans, and monitor performance against objectives and budgets. That support and guiding role seems to have broken down, as Oxfam has focused its attention more on the *Mahasangha*, hoping that, in the meanwhile, BOJBP would 'sort itself out'.

Tension between BOJBP and the Mahasangha

There is no doubt that the separation between BOJBP and the *Mahasangha* has been painful on both sides. We consider now in more detail the causes of this separation and the continuing tension between the two.

Different ways of working

While the one grew out of the other, there is a basic difference in the way the two organisations have developed. BOJBP is a voluntary organisation, registered as a society under the Societies Registration Act of 1860. A 'society' is a body of at least seven like-minded people who work to specific objectives without profit or commercialisation. It is managed democratically by those who form it, and, though it may seek wider membership, it is not required to do so. Because societies are registered with the government and can receive resources from it, they may be considered to be part of 'the system', somewhat constrained, therefore, from challenging it. People's organisations, on the other hand, are not registered and cannot apply for public funds. They have the potential to challenge the established order. They aim to be widely democratic, working from the mandate of a mass-based membership, and seeking to address the issues of their members through campaigning and lobbying. It is clear that, while BOJBP has always been a 'representative' organisation with wide participation of large numbers of people, it has never sought to have a mass-based membership. Essentially its leaders have always led from the front. But the *Mahasangha* now genuinely seeks to represent all 400 villages[5] that are federated within it, and to have them determine, through their elected representatives, the policy directions and strategies of the organisation.

These conceptual differences between the *Mahasangha* and BOJBP, which relate to fundamentally different ways of working, compound the other differences between them. The seeds of these other difficulties were sown when the *Mahasangha*, together with some BOJBP staff, moved – albeit at the behest of BOJBP – to Nayagarh town at the beginning of 1996. Significant though this move was, however, it was not the sole root of the problem. For at about the same time, as we have mentioned, the nine most active sister organisations were granted 'autonomy' by BOJBP. This meant that they no longer needed to report to BOJBP. Instead, a Co-ordination Committee was set up at the *Mahasangha* level to which they, along with their Field Organisers, would report. While this was done, as we stress, with the agreement of BOJBP, the full implications of it were not appreciated at the time. Perhaps so long as BOJBP felt that the *Mahasangha* was 'theirs', they believed that they would have ultimate control, even if they had lost it on a day-to-day basis.

'How can the mother become a sister?'

But there was another problem. It arose from the *Mahasangha's* attempts to make the sister organisations more democratic and transparent. Feeling that many of them had become relatively closed and undemocratic, towards the end of 1997 the *Mahasangha* started a process of making them more truly representative of the wider forest-protecting communities whom they served. A team from the *Mahasangha* visited each sister organisation and IVFPC in turn, to discuss with the people the importance of the process of democratisation and the purpose, structure, and functions of the *Mahasangha*. The ideas they put forward were widely welcomed. However, in their discussion with the Executive Committee of BOJBP they met strong opposition to any notion of restructuring and democratisation. The Committee preferred to function as a voluntary organisation with the existing membership from 22 villages, and saw no need to change. This attitude was an unpleasant surprise for people in the *Mahasangha*, who were simply unable to digest it. How could the organisation which had taught them so much about people's participation and voluntarism, and which had founded the *Mahasangha*, now oppose democratisation? It was simply beyond their understanding, and they lost respect for BOJBP.

While tension began to develop over this and other issues, so long as Jogi, as President of the *Mahasangha*, kept a foot in both camps, it was a tension that could be managed, at least for a while. But the *dénouement* could not be postponed indefinitely. As the *Mahasangha* built its relationships with the sister organisations and IVFPCs, and helped to strengthen them and make them more democratic; as it established links with other forest networks in Orissa; as it gained a voice on behalf of its members in many district and State-wide forums; as it built its own relationship with Oxfam; as it became a centre for exchanging information and ideas; as it became a place to which researchers came to study forest-protection activities; as it took on more staff and trained them; as it built its credibility — as all this developed, so BOJBP became gradually more and more marginalised. Finally, when midway through 1998 they saw that the *Mahasangha* had become a fully functioning entity, with its own management committee and mass-based membership, BOJBP felt that the organisation to which that it had given birth had taken over its own 'family' of sister organisations, including BOJBP itself! Speaking of this metamorphosis, one of them said to us: '*How can the mother become a sister?*'

Sadly, towards the end of 1998 BOJBP felt it had no option but to withdraw from the *Mahasangha* entirely, and refused to be the channel through which Oxfam funded the *Mahasangha* and the sister organisations. Perhaps what is saddest of all is that many of the key people in BOJBP still simply do not accept the *Mahasangha*. In spite of the fact that they are the ones who set it up and participated so fully in its early activities, they have been unable to move with it as it matured, and in the end could not 'let it go' graciously. Instead they have come to resent it, even actively disliking it, criticising it at every turn, and blaming it for many of their own difficulties.

All of this, of course, has distressed many in the *Mahasangha*, Jogi especially, who has been singled out for particular blame. He had worked tirelessly to keep BOJBP involved, to help it to open itself up, to become more democratic, and to understand the importance of the *Mahasangha* and support it. So great became the divide between the two organisations and BOJBP's hostility to the *Mahasangha* that he felt he could no longer be BOJBP's representative and President of the *Mahasangha*. The ideological and personal differences were simply too great to bear. He, therefore, resigned from both in July 1998.

Opinions on the way forward

We had many discussions with people, within both BOJBP and the *Mahasangha*, on the way forward for BOJBP. Here we summarise the main points that were made to us:

- Several mentioned the internal differences of opinion within the leadership, which have to be settled through open and honest discussion. '*The leaders have to be clean and genuine, otherwise the organisation will not survive*', said one. '*Conflict within the house has to be resolved.*'
- As we know, much of this internal conflict is centred on BOJBP's relationship with the *Mahasangha*, so that one directly affects the other. Until there is full agreement on what the relationship should be, then tension between the two will continue.

- Because of the conflict between BOJBP and the *Mahasangha*, there is a separation of the Women's Development Programme from the main programme, so that women are not integrated into mainstream environmental work. This is not true everywhere, but it pertains within BOJBP and some sister organisations. BOJBP is holding on to the Women's Development Programme, resisting what it sees as the threat of a 'take-over' by the *Mahasangha*. '*Men and women together have to protect the forest and environment*', said one person to us.

- Several people spoke of the need to remotivate and involve people in the 22 villages of the mother area. Acknowledging that there is dissension and apathy, one said to us, '*Unity and solidarity within the village [Kesharpur] and among all 22 villages have to be maintained ... and their participation has to be ensured in the movement and the programmes.*' For that to happen there needs to be clarity of purpose, '*so that the confusion in people relating to BOJBP is cleared up*'.

- Some spoke specifically of the need for the involvement of young people as an essential part of the revitalisation process. Young people played a key role in the early years. They need to be involved again now, in the organisational management as well as in the implementation of the programme.

- Recognising that the maturing forest has become a valuable resource that needs proper management, several people emphasised the importance of developing effective benefit-sharing mechanisms. '*Otherwise,*' warned one, '*the forest will not stand for long.*'

- The importance of strengthening village-level organisations (such as Village Forest Protection Committees) was stressed.

- And so was the clarification of roles: '*All office bearers and members of staff must be clear about their roles and responsibilities within BOJBP.*' For that to happen, there has to be '*a process of meetings and consultations*'.

- In the course of telling the story of BOJBP, we have pointed to the organisation's over-ambitious programme activities. This was recognised by many of our interviewees, and there was a plea for activities, once started, to be carried through to ensure 'continuity of programmes'.

- Finally, there was a strong appeal to return to the simple virtues that drove the organisation so successfully for many years. '*The challenge before us*', we were told by one person, '*is to sustain the interest and respect of the people for brotherhood and village unity. Only then we can succeed in our mission.*'

Another agreed: '*BOJBP should aim its efforts to ensure honesty and simplicity in its management, for which it is famous.*'

Summary and conclusion

It is clear that the second half of the 1990s has been the most difficult period in the life of the movement, and that BOJBP now stands at a cross-roads. The dynamism and creativity that were the hallmarks of its first ten years have dwindled, the leadership has lost it way, and the organisation has to some degree atrophied. And yet it has given birth to a remarkable offspring – the *Mahasangha* – which has picked up so many of the challenges that BOJBP itself first identified and engaged with. In so doing, it has made itself into a highly effective campaigning organisation, genuinely attempting to give voice to the people of Nayagarh and Khurda Districts involved in forest protection, successfully defending their interests against the predations of the 'wood-eating tigers' of Social Forestry, and resisting the seduction of JFM by the Forest Department. While striving to be a true people's movement, it has made its structures more democratic, its membership broad-based, its functioning more transparent, and its member organisations (sister organisations and IVFPCs) more accountable to their members. Whatever may be the current difficulties within BOJBP and between BOJBP and the *Mahasangha*, it is vital to acknowledge that the one would not have come about without the vision of the other. If, over the next year, BOJBP, with support from Oxfam, can tackle its immediate internal problems, and can come to terms with and accommodate the *Mahasangha* – even if it does not join it – there is no reason why its present difficulties should not rapidly become past difficulties. For this to happen, courage and effort will be needed, to admit to some uncomfortable home truths; to listen to the views of people such as those whom we have interviewed; to make peace with the *Mahasangha*, to learn from it and to draw strength again from the very people who made the organisation work with such effectiveness in the past: the people of the 22 villages.

The elements of success 7

'People are our strength.' – BOJBP slogan

Despite the many challenges that are currently facing BOJBP, its amazing achievements should not be forgotten. This chapter examines the factors that have contributed to those achievements.

High-quality leadership

The foremost factor is the quality of the leadership. While many dedicated people have been involved in the movement, clearly there have been three key people without whom little would have happened. First there was Udayanath Khatei (*'Bapa'*), the 'home-grown' leader: local small farmer, much respected by his fellow villagers for his sacrifice, service, and hard work, leading from within the community, bringing people with him. Second, the 'social activist leader', Joginath Sahoo (see Figure 13), respected locally as a teacher, inspiring people through his vision, commitment, and integrity. (*'Jogi babu is the real spirit behind the movement. Because of his sincerity, dedication, and sacrifice, the movement became successful and BOJBP could be formed and strengthened'*, according to Ajay Mohanty, a Youth Leader in Manapur.) Third, Dr Narayan Hazari, who first instigated the movement and played a role very similar to Jogi's in the early years, but later used his position to collaborate with researchers, academics, the NSS, other NGOs working on forest issues in and around Bhubaneswar, and people in the Orissa Forest Department. (*'Professor Hazari is the real motivator and key person in creating this movement. He educated the villagers on the impact of environmental degradation. He told his fellow villagers that our village would be completely smashed, as the stones would come rolling down from Binjhagiri hill'*, we were told by Susanta Jena, the first paid worker of BOJBP.)

All three leaders, however, would say that they could have done nothing without a second line of leaders, particularly school teachers and community leaders in other villages who helped to carry the people with them.

What are the special qualities of the leaders? Undoubtedly their simplicity, honesty, and sincerity of purpose have over the years attracted and motivated

Figure 13: Jogi (second from left) in campaigning mode. In the early years Jogi and his colleagues spent huge amounts of their time persuading people to join the movement, and until every family in every village had done so, the movement could not really take off.

many thousands of people in hundreds of villages. Here were people who could be completely trusted, who worked with total dedication, and who were manifestly not motivated by the desire for personal gain. Here were people, too, who offered hope in the face of impending ecological, social, and economic disaster, with a vision which could be realised in practice. Their approach through selfless voluntarism and the Gandhian principles of non-violence also had enormous appeal. They caught the imaginations of people through their powerful and often highly innovative techniques of mass mobilisation and persuasion. They have been very open, ready to listen to what ordinary people have to say, to learn from what others have written, to hear from other activists, and always ready to put their learning into practice. Ready, too, to offer warm friendship and hospitality to visitors and embrace them into the BOJBP family, as we have personally found. In addition to all this, they have been extremely skilful in dealing with external organisations and government agencies, always seeking co-operation in preference to confrontation.

It is also true that the leaders have demonstrated not only great empathy towards other people but also a profound and even spiritual love. If the notion of love as a development 'tool' jars with some readers, we can only say that it was the leaders of the *Mahasangha* who offered this to us as an explanation of the unique effectiveness of BOJBP's leadership: *'They really do love the people, and the people love them.'* Leadership of this quality is as rare in India as anywhere. However, it has a negative aspect, in that it has created an excessive dependency, particularly on Jogi, the consequences of which have become apparent in the last two years. Kant et al.[1] gave a warning about this in 1991, when they wrote:

> Once the people lose faith in BOJBP, the whole system may collapse.
> Another factor which may affect sustainability is the possible non-availability of leaders of Mr Joginath Sahoo's credence and devotion in the future.
> An organisation like BOJBP cannot work without strong leadership.

Villagers pulling together

Within most Indian villages there is a hierarchy of caste and class that divides rural societies into many layers. In certain parts of India, such as Uttar Pradesh and Rajasthan, caste hierarchy is rigid and extreme. Elsewhere it is less so. But nowhere in Hindu India is caste division not present. (It has even penetrated many tribal communities that have been 'sanskritised'.[2]) Within the 'mother area' of BOJBP, however, caste division is not great. Most of the 22 villages are populated largely by families from the *Khandait* caste, a lower-middle caste group. The vast majority of these families depend on their farms for a living; of those who do not, most are landless farm labourers. Furthermore, the pattern of land ownership is not extreme: few holdings are large. There are not, therefore, such extremes of wealth as one finds throughout much of rural India.

Undoubtedly the relatively small differentiation in caste, class, occupation, and wealth, both within and between villages, has contributed to the readiness with which communities here have worked together and have resolved their differences. Furthermore, in order to reduce the potential divisiveness, and the factional politics of caste, the leaders of BOJBP some years ago declared that, whatever caste people had hitherto belonged to, from now on they were all members of one caste, 'the Forest Caste'. By this means people came to see that their common purpose – of environmental regeneration and protection – transcended residual caste-based allegiances. Under the aegis of BOJBP they were all stakeholders in the project, regardless of social background.

It is also important to note the long-established tradition of communal action and voluntary labour in many villages in the area. Examples include building the clinic in Kalikaprasad, and the schools in Kesharpur. For some decades Kesharpur has prided itself on its particular spirit of unity and co-operation, for which it has given itself the name *Buddhagram*: 'the village of enlightenment'. However, while relative social homogeneity and a tradition of communal voluntary labour make co-operation easier and reduce the likelihood of conflict, they will not prevent conflict from arising. The recent problems within the organisation have occurred in spite of the relatively favourable social structures, democratic institutions, and well-established traditions of co-operation.

Benefits for all?

To succeed in its goal of forest regeneration and protection, the movement needs the co-operation of all members of the local community. From the very beginning, every family in all 22 villages agreed to restrain themselves from exploiting the forest and to abide by the rules of forest protection. For some people the costs were considerable. There was, for instance, the matter of giving up goats. This was a particularly severe sacrifice for those without land – the very poorest families – since they were heavily dependent on their goats for milk, meat, and cash. However, we understand that in spite of the agreement reached between the villages in 1982 to sacrifice their goat-keeping, it was only the people of Kesharpur who actually did it: the rest simply controlled where their goats grazed. Nevertheless, for the poorer people of Kesharpur it was a sacrifice that had to be recognised; and the community did indeed grant them compensation, in the form of money from the village fund to rent land from larger landowners.

The people in the six villages that had previously been entirely dependent on the forest of Binjhagiri have received no direct benefit from their self-restraint, and no attempt was made to compensate them for their loss of access to the forest on the hill. As a result they have been forced to buy fuelwood from Forest Department depots. But, while they have got nothing directly in return for their restraint, they have clearly benefited from an overall improvement in the local environment, through recharged springs, a rise in the water table, and a reduction in the impact of droughts.

Every villager in the nine hill-foot villages has rights to collect dried twigs and leaves from the forest, and to cut freely from certain bushes. In practice this right has benefited the poorest families the most, since the better-off homes would have sufficient trees on their own land from which to harvest fuelwood; or they could afford to buy it; or they would use agricultural waste from their fields. In addition to their rights to gather sticks and fallen branches, every family has rights to gather fodder, medicinal herbs, tubers, leafy vegetables, and fruits. Again, in practice, the benefit has been greater for the poorer families, with little or no land, than for the better-off, whose farms could provide for most of their needs and whose cash could satisfy the rest. Also, as the forest matured, more products became available, and the poor could sell some of them for cash. In this way the poorest could benefit both in terms of subsistence and in the form of cash. This was not because the system was designed to be weighted in their favour: it was simply how it worked out in practice.

In theory, everyone has had an equal chance to serve on the Village Councils, although in many cases members of better-off families have been chosen to 'add weight'. This has meant that in reality the better-off may have been over-represented, and the poorer sections under-represented. Furthermore, in those villages that did not operate Kesharpur's extremely democratic system of voting, there was the possibility of nepotism and corruption, which in all likelihood would benefit those already wielding power and influence. Nevertheless, it must be stressed that everyone has equal rights in voting and in representation.

To conclude, even though the system of rights and responsibilities has not been based on equity, in the sense that the poorest have not been especially favoured, overall in practice it has not worked to their *disadvantage*, and in some respects it has even worked to their advantage. They have been willing to support it, therefore, in the greater cause.

Winning friends and influencing people

Gandhian philosophy and the use of a range of non-violent techniques of persuasion have been extremely important tools in the development and success of the movement. Personal appeals, prostrations, *padayatras*, and fasting have all played their powerful parts on many critical occasions. The leaders have astutely cultivated relationships with foresters and other government functionaries, a matter to which we return later.

At the same time, they have striven to solve their own problems in their own ways and to avoid involving outsiders in resolving disputes. This has included dealing with miscreants accused of stealing from the forest. To these they have usually meted out appropriate justice in the form of requiring apologies and community-service work (planting trees), rather than through the imposition of fines, an approach which has engendered goodwill rather than resentment.

The leaders have been enormously successful in building a sense of solidarity and unity within the communities, to the extent that, when it has been necessary to mobilise people for a protest or a *padayatra*, they have been able to do it with ease, sometimes gathering hundreds of people together very quickly. Theirs has always been an inclusive approach, embracing all castes and conditions of people. The result has been a strong and wide sense of ownership, both of the organisation and of the environment. One of their slogans has always been, *'People are our strength'*.

They have also addressed many of the basic needs of the people and the issues facing them, including the need for fire-prevention initiatives; the establishment of income-generation schemes and thrift and credit groups, specifically for women; the resolution of age-old conflicts; the challenges of untouchability and dowry; the provision of adequate sanitation, and other health-related matters, such as leprosy.

Finally, as well as developing dramas, songs, and slogans, they have used a range of printed materials, posters, leaflets, postcards, and booklets to persuade people of the importance of the cause and to help them in many practical ways to engage with it.

Working with children and working through children

School children have been crucial to the development of the movement in several ways. First, they have acted as unofficial but highly effective ambassadors for the environment, spreading the messages of protection and regeneration within their families and wider communities. They have been equipped for this role through the opportunities seized by their teachers in the formal curriculum. While teaching methods are quite traditional, many imaginative and innovative techniques have been used to awaken and sustain their interest, such as writing letters to trees, acting out their own plays (Figure 14), and composing 'green songs'. Awareness has also been fostered through extra-curricular activities, such as 'green clubs'; environmental quizzes and inter-school debating competitions; and school seedling nurseries.

Second, children have been very actively involved in plantation work on the two hills, and in the compounds of their schools and homes. *'I have planted ten saplings at my home, and eight of them are surviving'*, we were told by 12-year-old Madhusudhan Dora, who is in the Middle School in Kesharpur. *'These are guava, mango, and other fruits. I also planted five trees on Binjhagiri two to three months ago.'* Other children gave us similar reports, all with a clear appreciation of the benefits of forest regeneration. Said Bisnuprasad Sahoo, another 12-year-old from Kesharpur, *'Sometimes I go into the forest with other children and collect berries. Because of the forest we are getting oxygen, good rain and wood.'*

The elements of success

Third, children have been active in campaigning on the environment, through *padayatras*, dramas, and other cultural events, and have also been rallied in defence of the environment when it has been under threat. Many children told us of their participation in awareness-raising events, and others told us of their involvement in forest protection. Gokula Maharana, a 14-year-old at Gambhardihi High School, related with some pride, '*We have even caught woodcutters in the forest!*'

Using children as the agents of campaigning and protest is radical, and would perhaps be questioned within the context of many societies, both in the North and South. But the leaders of BOJBP, Joginath Sahoo especially, have always seen children as legitimate and highly effective agents for change. Inculcating awareness and commitment in the young has been seen as a way of helping to secure the future. There has always been the hope that the child activists of today will become the adult activists of tomorrow. And in the history of BOJBP there are many cases where this has happened.

In spite of all the current difficulties, the schools programme continues. Children continue to plant trees, tend nurseries, hold green debates, and raise awareness in their families and communities of environmental issues.

Figure 14: Children acting out a play which teaches the importance of protecting animals as well as trees. The involvement of children in the movement has been one of its great strengths, marking it out from many other environmental movements. Children have not only been used to convey the messages of environmentalism from school back to their families and communities, but they have also been used to campaign against attacks on local forests.

133

Spreading a 'green culture' and tapping into a 'green spirituality'

In the minds of the leaders in BOJBP, the work of the movement has always meant more than simply planting and protecting trees. Theirs has always been a wider vision, one that has a moral and spiritual dimension. In Chapter 3 we noted the closeness of the relationship that binds tribal communities to their forests, and we described the ways in which that relationship is maintained through certain practices and rules, and is expressed in their myths and legends. For the leaders of BOJBP, at least, their regenerated and protected forests have always deserved the same deep respect accorded to them by tribal societies. They have tried to engender this respect principally by making trees central to many community activities. No festival or religious ceremony takes place without the planting of trees to mark the occasion, and many social/religious ceremonies, such as marriages and funerals, are marked by tree planting. Saplings are used as gifts and as prizes. Children come home from school and plant trees around their houses and on the hills. All visitors to BOJBP are expected to plant trees on the campus before they leave (Figure 15). Tree planting is commonplace in the life of the community, something which people do habitually.

In their greetings, local people quote, chant, or sing slogans about trees. '*We can't live without trees*' is an oft-repeated slogan in everyday life. By using

Figure 15: Visitors are *expected* to plant trees. This is not just a commemorative act but also an act of solidarity, binding the visitor into the movement. This photograph was taken in 1991; the tree is now 15 feet tall, and the visitor (Joe Human) receives regular news of its progress.

expressions about trees – one might even call it a 'language of trees' – people's awareness is constantly reinforced, as is their sense of belonging to a 'Forest Caste'. Meetings, workshops, and seminars commonly start and finish with a chant about trees, and in the course of *padayatras* chants are sung all the way. Religious and moral discourses are often used to explain the importance of the spiritual and moral imperative of environmental awareness and action.

Nevertheless, one must question the extent to which a 'green culture' and a 'green spirituality' have really taken root among ordinary farming families. In the view of several observers, the leadership of the organisation has taken greenness into its heart and soul, as have many of the school teachers and children active in the movement. But what about the wider community? These are not forest-dwelling communities. They have some dependence on the forest, but they are unlike tribal communities in that they do not have a strong sense of identification with the forests. They are settled farmers, for whom the forest is important, but not integral to their cultural, moral, and religious frame of reference. Nevertheless, the regenerated forests of Binjhagiri and Malatigiri stand as witnesses to the efforts of every villager in 22 villages who has taken the message of environmentalism to heart and responded, at the very least, by not destroying the forest any further and, at best, by participating actively in its regeneration and protection.

Links with external institutions

We have shown the importance of Dr Narayan Hazari in making links with external agents. Over the years many of these became important to the movement. First there were people in academia, researchers and others working on forest and environmental issues. These he pointed in the direction of Kesharpur, so that they could visit and learn about community-based environmental activism, in return for which they were expected to lend support and give credibility to the movement in a wider context, and where possible contribute from their own experience and perspectives.

Then there were government officials, for whom Dr Hazari considered exposure to the work of BOJBP was important. These included people from the State Forest and Environment Departments. At one time the Forest Department regularly sent trainee foresters to Kesharpur for exposure visits as part of their training. Even if such official visitors did not end up on the side of the organisation, at least they were less likely to be against it. Dr Hazari also facilitated visits by students from the National Service Scheme at Utkal University to Kesharpur to participate in replanting programmes, and to show solidarity with the people.

Publicity has also been gained by many articles in journals and newspapers, and chapters in books that Dr Hazari and his brother, Subas, wrote on the environmental activism of BOJBP. These served to spread knowledge and understanding of the movement to wider audiences, both within Orissa and beyond, in India as a whole.

The role of Oxfam

Of the other external agencies, Oxfam must clearly rate highly. Oxfam's relationship with BOJBP is an exceptionally long one, beginning in 1984,[3] when it made its first grant to assist people in the fire-damaged village of Awasthapada. Throughout the intervening years Oxfam has been the sole non-government funding agency, by the choice of BOJBP itself. The support given by Oxfam in monetary terms is very small: in the first ten years it amounted to a little over £30,000,[4] including two emergency grants. But Oxfam's role in providing non-funding programme support has been as important as the money itself. Its contribution includes information sharing, support with planning, help in working towards gender equity, capacity building, learning from others in the same field, alliance building and networking, and providing links with the outside world. Jogi says that it is *'Oxfam's philosophy and ways of working that have attracted us towards Oxfam over other development support agencies.'*

Needless to say, Oxfam has benefited a great deal in return. It has taken many dozens of visitors to Kesharpur to learn from BOJBP and to apply the lessons more widely. While hosting visitors has consumed valuable time, their visits have given support and encouragement to the organisation. Many of these visitors have also been instrumental in spreading the message widely in the United Kingdom, through illustrated talks and articles in magazines and newspapers. In this way the work of BOJBP has been brought to the attention of thousands of people, many of them Oxfam volunteers and sympathetic members of the public, who have given their commitment and financial support to Oxfam. There is also the *Thengapalli* Development Education Project, which has reached tens of thousands of primary-school children in Britain (described in the Appendix). From all this, it is clear that the relationship between Oxfam and BOJBP has been one of rich mutual benefit.

Finally, with its State-wide view of development issues in Orissa, particularly those relating to forests, Oxfam has played a significant role in scaling up the work of BOJBP through the formation of the *Mahasangha*. While this has caused problems for BOJBP, it is highly significant, as seen in Chapter 6.

Other key players
Other important external supporters of the movement are the agencies that have given it awards. Of these, the international UNEP Global 500 Award was the most prestigious, and considerably enhanced the reputation of the organisation. A significant visitor, who gave an enormous boost to the morale of BOJBP's activists in 1983, was Sunderlal Bahuguna, the nationally revered and internationally renowned environmental campaigner. The visit of someone of his reputation demonstrated to many local people that what they were doing had wider significance and resonance.

While the relationship with the Forest Department has not always been satisfactory, during the early years BOJBP worked with their staff to regenerate the forests on the hills of the area, and then again during the era of Social Forestry planting. We mentioned the valuable support of the Divisional Forest Officer, Pratap Patnaik. There were others whose support was important, if not vital. Over the years there have been many disputes with foresters at the local level, but these should be set in a context in which for some time relationships – at least with senior staff – were good. The story is different now, as we discuss in the next section.

The significance of Joint Forest Management
One other factor which gave strength to the movement, in a quite unexpected way, was the JFM declaration issued by the Orissa State government in July 1993, which had the effect of changing the status and reputation of BOJBP and the *Mahasangha*. Until this time they had been essentially promoters of community-based forest management (CFM) at the local level; but the government's JFM declaration turned them into high-profile campaigning and lobbying agencies, raising questions about the rights of local communities over forests within the new JFM framework. As we have seen, their approach to JFM is basically hostile, and so is the attitude of many local communities involved in CFM work. BOJBP–*Mahasangha*, therefore, became the champions of local communities, constituting an organisation with which forest-protecting communities could identify, which could be trusted to fight for them and *their* rights. With the huge support of a large number of villages, BOJBP–*Mahasangha* have been able to build a network to establish the rights of forest-protecting communities in the face of JFM, one much larger than they had before. Recently this new strength and support has enabled BOJBP–*Mahasangha* to play a pioneering role in the formation of the *Odisha Jangal Manch*, the State-level forum of forest-protecting communities.

Summary

We have shown that several factors have been responsible for welding BOJBP and the *Mahasangha* into an effective organ for the promotion and defence of community-based forest-protection work. Strong, very committed, and imaginative leadership has been a key ingredient. But that leadership has been enabled to work by the reciprocated trust and commitment of the people. Communities have been enabled to pull together by their relatively homogeneous social structures. In many villages there has also been a strong tradition of communal work. Although BOJBP has made no attempt to redistribute resources, nevertheless the benefits accruing from forest regeneration have been shared among all sections of society. The activists have been astute in the ways in which they have won people over, both within the communities involved and outside them. They have also been far-sighted in their involvement of children. The extent to which a 'green culture' and a 'green spirituality' have been generated or activated remains questionable, but at the very least many of the leaders and activists have internalised deeper moral and spiritual values. Oxfam has clearly been a key external player, providing both financial and non-funding support. A number of academics, distinguished visitors, and foresters have also lent support at critical times, and students from the NSS played a supportive role in the early years. Each award, of which there have been several, gave the organisation – particularly the leadership – a boost. Finally, JFM has brought many other organisations into the BOJBP–*Mahasangha* family and given it a reputation as an effective defender of community-based forest protection and management.

Lessons learned   8

This spontaneous and creative movement brought through local initiative has given meaning and purpose to the life of many men and women in the villages. They have derived a sense of fulfilment from this and will continue to draw inspiration from this on-going programme.
(Narayan Hazari and Subas Chandra Hazari[1])

Among Indian environmental movements there is no doubting the astonishing achievements of BOJBP – whatever its current difficulties. Through it, and more recently through the work of the *Mahasangha,* more than 500 villages in the districts of Nayagarh and Khurda have been brought together on a common platform of community-based forest management, and have achieved the protection of more than 50,000 ha of forest. The BOJBP–*Mahasangha* family of sister organisations and IVFPCs has also successfully implemented the Government's Social Forestry Programme, with hundreds of communities raising village wood-lots and ultimately protecting them from the predations of the pulp and paper industries. The waves of the movement have spread far beyond Nayagarh and Khurda Districts to other parts of Orissa, motivating villages to take up community-based forest protection and to federate for collective action. As we have shown, they have also rippled far beyond Orissa, to other parts of India and abroad. From all these achievements, there are many lessons to be learned.

Lessons for policy makers

The first lesson must surely be that forest-dependent communities themselves have the capacity to undertake forest regeneration, protection, and management without the assistance or intervention of the Forest Department. They can also do it considerably cheaper, because they rely mostly on voluntary labour. Long before the advent of Social Forestry, many communities were doing their own plantation work, using stock from their own nurseries and their own labour. Should anyone doubt the scale of their achievements, they have only to consider the extent of regenerated and replanted forest now clothing the hills of the area. For years the accepted wisdom has been that only the Forest Department has the resources and expertise to look after India's forests. What has happened in Nayagarh, Khurda, and elsewhere clearly gives the lie to that. Here communities have achieved a sense of ownership over their local forests and forest produce. They have formulated rules for forest protection and usage. While they may not yet have taken into account the needs of special-interest groups, in most cases they have ensured that no one loses out. In some places effective systems of forest management and usufruct sharing are emerging (for example, *Sulia Paribesh Parishad*), and the *Mahasangha* itself is starting work on this important matter. With some exceptions they have also been able to address conflicts within and between villages, without recourse to external mediation – the Forest Department, the police, or the courts. They have developed effective strategies for raising people's awareness of forest conservation and environmental protection; for establishing and supporting new community forest-protection organisations and networking with them; for learning from best practice and developing new ways of working; and for developing better policy analysis and appropriate advocacy responses. There is much here for policy makers in the government and the Forest Department to consider.

Lessons for bilateral donors

We are here considering the major donors involved in supporting Indian forestry, such as the World Bank, SIDA (Swedish International Development Authority), and DfID (Department for International Development of the UK government).

There needs to be a meeting of minds between the so-called 'experts' who practise scientific forest management and the people who practise community-based forest management, and the big donors must be instrumental in organising this. The experts need to be helped to demystify their subject and make it more widely accessible, at the same time as recognising and utilising the knowledge and skills of forest communities, particularly those pertaining to the conservation of biodiversity and to silviculture. The balance between forest growth through plantations and forest growth through natural regeneration is presently tilted hugely towards the former, with few resources devoted to the latter. This distortion should be rectified, with much more attention being paid and resources allocated to the regeneration of native forests, and less to the plantation of exotic forests. An appropriate intervention here could be technical support for the raising of nurseries and the management of seed banks to generate the stock for native plantations, learning from the BOJBP experience. The forest-management responsibilities of the government, on the one hand, and of local people on the other need to be developed with full community participation in a relationship which values genuine partnership and respects local knowledge and experience, rather than one in which communities are ignored and their knowledge disregarded.

Accompanying this there needs to be a complementary shift in resources allocated to forest-protecting communities and the organisations to which they belong, such as BOJBP's sister organisations and federations like the *Mahasangha*. Bilateral donors should help to build the capacity of these entities, to support them in securing greater rights over forests and forest produce. At the same time, they ought to influence the government to allow greater freedom to forest-protecting communities to protect and manage their forests. It is clear that inter-community co-operation and networking has been crucial to the success of BOJBP's approach. Bilateral donors must accept that networking is an appropriate activity to support, to help to strengthen communities and their federations. In addition they should lobby the government for formal recognition of these networks and their initiatives.

As we have seen, systematic forest management was never a priority for BOJBP. From its birth, its aim was simply to revive and protect the degraded forests, not to manage them for sustainable usage. For them 'management' – if it meant anything at all – meant (a) formulating and implementing rules for forest protection, and (b) ensuring that people took from the forest only what they needed, and in the process did no harm to the forests. In their support for participatory forest management, bilateral donors should help communities to develop proper forest-management systems which take into account, on the one hand, the capacity of the forests to produce timber and non-timber forest products and, on the other, the needs and demands of the forest-protecting communities. Their support should also be sought to help forest communities to enhance their skills not just in the collection of NTFP, but also its production, processing, and marketing.

As the forests have matured, it has become apparent that a purely self-regulatory approach to forest 'management' (taking from the forest only what one needs) is inadequate, as neighbouring communities, vying for their 'share' of the forest, have come into conflict with each other. Bilateral donors could offer help with conflict management and resolution, where skills of rule formulation, observation of procedures, and documentation are needed for those bodies playing a judiciary role. They could also help with 'process documentation', to enable forest-protecting groups to report on progress against objectives (monitoring), so that their learning can be shared with others.

Although, as we have suggested, there are many areas of assistance open to bilateral donor intervention, it is vital to stress that community forest management is by and large a self-resourcing, self-sustaining process. Too much external intervention can be very destructive. Donors must resist the temptation to provide big funds for big schemes within which community identity and spirit are lost.

Lessons for Oxfam

Oxfam is the only non-government agency to support BOJBP. Having played this role since 1984, it can claim much credit for promoting and supporting the concept of community forest management in Orissa. However, it is also in part responsible for some of the current problems facing the organisation. If these are to be resolved, Oxfam must be actively engaged in helping to find solutions.

BOJBP has long been characterised by a heavy dependence on a fairly small number of key functionaries, centred on Joginath Sahoo, Udayanath Katei, and Dr Narayan Hazari. In the early years, as the organisation grew, so their capacity grew with it. However, there was bound to come a time when they needed critical support to help them to develop further their organisational and programme skills, and to develop the capacity of those around them. That time was first identified by an Oxfam staff member in 1988, and yet it took until 1993 before the first review was undertaken in which issues of growth and capacity were properly identified. In the same year a joint Oxfam–ODA evaluation recommended a number of changes and improvements, many of which were implemented. Since then there has been no further review, although at the time of writing one is imminent. In the face of a number of problems which we have identified, what can and should Oxfam now do?

The most immediate problem is the unresolved tension within the leadership of BOJBP concerning its relationship with the *Mahasangha*. Difficult though this will be, Oxfam must play a role in helping to resolve this problem, which is like a festering wound, affecting both the internal workings of BOJBP and its external relationships with the *Mahasangha*, with sister organisations, and IVFPCs, and with Oxfam. There is a very real danger that as long as the problem remains unresolved, the united front that is needed on JFM in particular will weaken and allow the Forest Department to negotiate with individual villages. It should be acknowledged that Oxfam's role in helping to mend fences is made more difficult because the agency is also part of the problem, being perceived by some to have favoured the *Mahasangha* over BOJBP. Openness and honesty will be necessary on all sides.

A second way in which Oxfam could and should help is in capacity building. Assistance is required on many fronts. To begin with, the leaders need much help in developing their skills to manage the organisation and the programme more effectively. This should cover the management and development of human resources, financial management, organisational development, and participatory programme planning and implementation. They need support too in the practical aspects of continuous field-based monitoring and regular evaluation. This will help to ensure that strategies are continuously evolving, that appropriate interventions are made in the face of new challenges, and that problems are identified before they become unmanageable. Emphasis should be placed (a) on organisational renewal and development that empowers new leadership, (b) on participatory governance, and (c) on effective and open communication within the organisation.

Support is needed from Oxfam in exploring and evaluating the range of options available for effective community-based forest-management systems. What systems will promote the most effective resolution of conflicts, and what will offer the most equitable usufruct sharing which takes into account the needs of special user groups: women, artisans, the landless, and *dalits*? These issues must be explored. Assistance is also needed to ensure that these groups can participate fully in the whole process of programme planning. Of critical importance is the mainstreaming of gender equity into BOJBP's management and programming.

Finally Oxfam must look at how the lessons learned from the whole BOJBP–*Mahasangha* experience – both the positive and the negative – can be used to promote community-based forest management in a more systematic manner in other parts of Orissa and India.

Lessons for other forest-protecting communities and NGOs

Chapter 7 examined the factors responsible for the success of BOJBP. We do not suggest uncritical imitation, but we do believe that there is much to be learned by other grassroots environmental groups from the BOJBP experience. There is first the matter of leadership, which has carried people with it through its openness, commitment, integrity, and sincerity of purpose. The leaders' espousal of Gandhian values of non-violence; their use of powerful and innovative techniques of persuasion; their involvement of people from across all communities; their spirit and practice of voluntarism; their openness to new ideas; and the trust that they have built up among ordinary people have all been hugely significant in winning people to the cause and mobilising them for effective mass action. While the organisation has never tried to be a mass-based people's organisation in the way that the *Mahasangha* is now aiming to be, it has always sought to be democratic and inclusive, apportioning rights and responsibilities equally to all.

BOJBP's attempts to get people to adopt 'green values', to develop a 'green culture', and tap into a 'green spirituality' may not have had deep impact, but its work with teachers to raise environmental consciousness and stir environmental activism among school children has been enormously important and could be widely emulated. We know of no other organisation in which the potential of teachers and school children has been so developed. Similarly, the involvement of enthusiastic student activists from the NSS in awareness raising and plantation work has played an important catalytic role, particularly in the early years. Others could learn from this too.

Finally, there is much to be learned from the process by which BOJBP expanded its area of operation through the establishment of sister organisations and IVFPCs, and the support that it gave to them once they had been set up. Much, too, is to be learned from the creation of the *Mahasangha* as an overarching federal body, and its attempts now to develop itself into a genuine people's organisation.

Summary

At a time when there are enormous pressures on Forest Departments to implement JFM agreements with individual communities, policy makers need to stand back and take a hard look at what they are trying to achieve and what has already been achieved by thousands of forest-protecting villages through community-based systems of protection and management. Bilateral donors can play a part in bringing the two sides together; in funding initiatives; in offering technical support to promote the regeneration of indigenous forests; in supporting networking; and in developing effective forest-management and conflict-resolution systems. They need to do all this without swamping initiatives with vast sums of money. Oxfam must accept its responsibility to help BOJBP to resolve its current problems; to build its capacity; and to support a process of renewal and participatory governance. It must also assist BOJBP in determining the most appropriate system of forest management: one which balances the needs of special user groups with available forest resources. There is also much that can be learned by other NGOs and forest-protecting communities from the whole BOJBP–*Mahasangha* experience: lessons about leadership, values, techniques for awareness raising, the involvement of children and students, and the propagation of sister organisations.

Appendix 1: *Thengapalli*: a resource for primary schools

'Over the past few years I have had the privilege of sharing the story of the people of Kesharpur with a very wide variety of pupils in Hampshire, and without exception children have engaged with this story of hope in a far deeper and more personal way than any other curriculum resource I have ever encountered. As the children explore the hopes, fears, anxieties, and successes through drama, they subconsciously invest a great deal of themselves, and as a result, the unfolding narrative of Binjhagiri becomes the unfolding narrative of each of their own localities, struggles, and beliefs.'
(Chris Kilby, teacher at The Cedars School, Nursling, Hampshire, England)

The seeds of an idea

In early 1992, following a visit to BOJBP, Joe Human gave a talk about the movement to a group of Oxfam Education Workers. Afterwards one of them, Dylan Theodore, approached Joe with the idea of using the story of BOJBP as a vehicle for raising environmental issues with children in Britain. In the months that followed, the idea was worked on by Dylan, Joe, and Pam Liddicoat, a teacher seconded to Hampshire Development Education Centre (DEC) in Winchester. Out of these discussions eventually emerged *Thengapalli*, a cross-curriculum primary-school education pack for Key Stage 2 of the National Curriculum,[1] which is the subject of this appendix.

Recalling the time when he first heard the story and saw its potential as an educational resource, Dylan told us:

> It was invigorating to have such an inspirational story and to feel some kind of link with overseas for the first time. The fact that children were so much involved, the fact that it had this positive message, the fact that this was an idea that we could learn from ... made me feel that this was the antidote to those [other teaching] packs ... which were just cold snapshots of life wherever it might be.

Following this first encounter, it took two years for the project to get started. The use of BOJBP as the inspiration for the project had to be agreed; funding had to be secured; the working relationship between the Project Coordinator and Hampshire DEC, where the project would be based, had to be worked out; and schools who were prepared to pilot the materials had to be found. The support of advisers in Hampshire Local Education Authority (the LEA) was also needed. The plan was to develop the project over a period of three years, starting in September 1993. In the event the whole schedule slipped by one year.

Getting started

Dylan's initial aim was to use the medium of drama to convey and learn from the work and messages of BOJBP. The intention was to appoint a good teacher to run it, with advice and support from Dylan and Oxfam. However, as things turned out, Dylan was to be much more heavily involved than he had foreseen. A review of Oxfam's Education Department signalled the demise of the smaller DECs, such as his in Southampton. He therefore decided to leave and he applied (successfully) for the post of *Thengapalli* Project Co-ordinator.

The project began in September 1994, with major funding from the European Union, and the rest from the Local Education Authority (LEA), Oxfam, and Christian Aid. From the start it was essential to get a core group of schools and teachers involved, so that ideas and materials could be developed and tested. A small conference in November 1994 produced the nucleus of such a group. However, the members warned Dylan that some teachers would be intimidated by the 'futures' dimension that he wanted to build into the project in accordance with the ideals of Agenda 21.[2] They also advised that, with a narrowing down of the National Curriculum, and less time now available for such subjects as drama, the project must also contain a 'locality study' that could be used in geography classes. This was the first step towards broadening the scope of the project.

How it grew

Within a very short time other components were added, as the result of a series of meetings with advisers and inspectors in the LEA, covering the subject areas of English, art, religious education, and music. This sudden enlargement of the project in terms of both its content and its vision was as welcome as it was surprising. But it posed a considerable challenge, and moving the project fowards as it grew in breadth and vision was not easy. More and more teachers were warming to the idea, but many lacked the confidence to try things out for themselves. Dylan thus found himself going into schools and demonstrating the use of the materials. While he needed to do some of this to try out ideas, he also needed feedback from practitioners who were managing without him.

Appendix 1

When it began, the project was intended for use in both upper-primary and lower-secondary schools. But it soon became apparent that its integrated approach would be impossible to apply at the secondary level, because of early subject specialisation, and that the project would, therefore, have to be pitched at upper-primary Key Stage 2 only. It also became obvious that there would have to be a study visit to Kesharpur, to gather materials and engage with the movement at first hand. In October 1995 Dylan Theodore, two LEA advisers, and a local teacher travelled to India, where they were joined by a photographer, Rajendra Shaw, well known to Oxfam for the quality of his work.

The importance of the study visit
The study visit was enormously successful, and the group returned with a mine of materials on which to draw from their very rich experience, as they developed the resources and took the project forward. They also had a superb collection of photographs by Rajendra which were to become an important part of the teaching pack when it was finally produced. The visit helped the group to understand much more about the context of BOJBP's work, to appreciate more fully its achievements, and to understand its dynamics and the roles of the key players. We asked Dylan to explain what the experience had meant to him.

> One thing was the daily life in an Indian village ... just realising how tough it actually is. (You can over-romanticise things if you are not careful.) Realising this then led ... to the question: "Who is BOJBP? Who is involved?" What seems to come out of people's testimonies, and from just watching, is that in terms of re-afforestation, farmers and their families would describe themselves as members of BOJBP. But in terms of cultural programmes and children's involvement, it is essentially teacher-led, with teachers' families and students and staff at the headquarters of BOJBP ... I found it quite difficult when I got back, because I realised that every story is so much more complex, and it took me a while to consider what to do: try to reflect reality, or concentrate on history.

In the end it was the memory of a remark made by Rajendra Shaw which helped Dylan to sort things out. Rajendra had said that whatever they – the visitors – felt about cultural activities being 'staged' for them by an elite group of teachers and students, they must not forget the achievements of the whole communities in re-clothing their hills in forest: '*Remember, you are in India. If you can move from A to B in thirty years like that, that is a stunning achievement. I see development projects all the time, and that* [pointing to the hill]

147

is amazing, I promise you.' It was clear to Dylan that this extraordinary story had to be told, together with aspects of the cultural context.

Naming, writing, and publishing

Just before the study visit, the matter of the name of the project came to the fore, as it was obvious that the original title, 'Education for Sustainability: Learning through Drama', was no longer appropriate: the project was not now exclusively about sustainability, and the medium was no longer just drama. So *Thengapalli* it became. Dylan explained:

> ❦ Remember the *thenga*, the stick or staff, used in patrolling the forest? Well, that stick represents both authority and responsibility: authority to protect the forest and shared responsibility for it. And with the passing of the *thenga* from one family to another at the end of each day passes authority and responsibility for caring for the environment. Figuratively – or metaphorically – speaking, you, Joe, gave me a *thenga* when you first told the story in 1992. I picked it up and am now passing on to teachers, and teachers will pass it on to their children, and those children will pass it on in turn. In that way the seeds of an idea dispersed from far away will hopefully germinate and bear fruit closer to home. ❦

Despite early misgivings about the name, the notion of passing on responsibility was immensely appealing. And so *Thengapalli* it became.

Writing began in earnest after the visit. The project had now become very large, and although 18 months of funding still remained, there was much to be done. One problem that hampered the writing was the growing demand from schools – and there were now 20 or so involved in the project – for more materials to work on. This impeded progress for Dylan, and in the end he had to produce a 'do-it-yourself' starter pack for schools to work with.

Another challenge was finding a publisher. Initially he hoped that Oxfam's Education Department would take it on, but *Thengapalli* did not fit with its publishing plans. After further frustrations with another publisher, Dylan approached Hampshire LEA Publications Group, some of whose members were advisers and inspectors already backing and writing material for the project. The Group was very sympathetic and offered to put up £20,000. The only disadvantage was the fact that the LEA was not a commercial organisation and did not, therefore, have the resources to promote the material commercially. But the advantage was that all members of the Group were educationalists who understood what they were taking on, and could see its potential as a major primary-school educational resource.

Appendix 1

Other players, the launch, and beyond

Besides the involvement of teachers, advisers, and inspectors within the LEA, a providential discovery was that of Sushmita Pati, a woman from Orissa who lives in Hampshire and who is expert in Oriya dancing. She encouraged children in several schools to try creative movement in Oriya style. There was also invaluable support from Oxfam's Asia Desk and from Oxfam staff in Bhubaneswar, who enabled the study visit to happen. But perhaps most important of all were the hospitality and generous affection with which the five visitors were received in Kesharpur on their study visit and which made the project come alive.

We asked Dylan if there had been any other 'big thrills' in the life of the project. He replied:

> ❮ Yes, the first time I saw the performance of *Thengapalli* at Nightingale Junior School, the first chords of the songs that we had been writing and slaving over, and a character called "Bapa", played by a ten-year-old boy, stood up there describing Binjhagiri. That moment! My ten-year-old daughter told me afterwards that she had a lump in her throat. That was quite emotional for me too. ❯

The second highlight was the visit to Britain of Joginath Sahoo and his colleague Purna Chandra Mohapatra, who came in May 1997 to launch the pack at the Commonwealth Institute in London. It was a wonderful and moving occasion when children from Fryern Junior School performed the drama of *Thengapalli*, Sushmita danced, and the story was told again. With this event it seemed somehow that a certain symmetry had been achieved. A story of inspiration *there* had led to an inspired educational experience *here*. The *thenga* had been passed on, across the continents and across cultures.

Some 20 months after that occasion, we asked Dylan if there had been any disappointments for him in the way *Thengapalli* had developed. This was his reply:

> ❮ Just one thing, I think, which is that I had always to some degree seen *Thengapalli* as a tool within Agenda 21, so community involvement – going beyond the school gates – would be important. It is a disappointment that most schools don't feel they can do that yet. Even if it's a wonderful topic, it's still a topic within the classroom. In 1992 [the year of the Rio Earth Summit] you had this feeling that people were going to start looking at the environment within the

community and that there would be a process going on, leaving the car at home and all the rest of it – and it hasn't happened. The opposite has happened, probably. We have all become more single-minded about our own comforts. In any case, in formal education the political mantra of "raising standards" frequently translates into a dilution of context and meaning for children.*

Whatever one's opinion of this, there is good news too. The current Curriculum Review being conducted by the Department for Education and Employment places emphasis on community involvement, on 'active citizenship' in all its dimensions, which implies looking beyond the school gates to the external environment – physical, cultural, moral, and spiritual. Dylan believes that *Thengapalli* will be well placed to support that.

To date nearly 800 *Thengapalli* packs have been bought by schools, and we hazard a guess that it has reached 40,000 children or more. Currently work is in progress on a booklet showing how the pack can be used as a resource for the daily Literacy Hour that every school must implement. In addition, a video of BOJBP and village life in Kesharpur has just been completed to complement the pack. There is also a *Thengapalli* website, and Dylan is considering a schools-linking programme.

Responses

We were curious to know what makes the difference between a school that is using *Thengapalli* well and one that is not. Dylan commented:

> *If the teacher is approaching India in an open-minded way, in an empathetic way, if they see that issues and values are as important as ... the delivery of knowledge and skills, then it works ... Good teachers "wind up" children's attitudes, breaking down preconceptions, looking at what is important in environmental issues, or in place study and so on, involving them, empowering them. Once you have wound that up, the quality of work that uncoils is much higher. If you have the attitude right – and attitude will come from the teacher – then you get the engagement of children.*

With the publication of the Literacy Hour booklet and the launch of the video, Dylan's work on the production side of *Thengapalli* will cease, but he will continue to offer workshops for teachers who wish to use the material, and he is doing that on a free-lance/consultancy basis. What he has been unable to do – and it is with the wisdom of hindsight that he regrets this – has been to evaluate the use of the pack. A fourth year of funding would have been required for this, but that was not included in the funding proposal.

However well or otherwise *Thengapalli* is being used, what do teachers and reviewers think of it? Chris Kilby, who assisted in the piloting of some of the materials in the pack, observes:

> ❝ As children have become aware of the notion that change begins with individuals, they have felt empowered and have acted to instigate changes that they feel strongly about. I can recall one girl, an eight-year-old child, who, after studying *Thengapalli*, decided she would raise people's awareness of blood sports. She wrote a heartfelt letter to her local MP, and also informed the press that she had done so. She was consequently interviewed, and her story and the minister's response were published. She had achieved something real, got something done, and thousands read her article. ❞

A reviewer in the *Times Educational Supplement*[3] wrote: 'The pack is a cornucopia of high quality resources ... Many teachers will want to find a place for *Thengapalli* in the new curriculum ... They and their pupils will benefit from doing so.'

One of us was able to attend a workshop[4] at which the pack was introduced to a group of teachers, who afterwards made these comments in an evaluation: 'His inspirational story and pack need much more circulation and publicity.' ... It's much better than other packs I have and for relatively little outlay.'[5] ... 'I wonder at the power of the story.' ... 'The pack emphasised how much we have to learn from other countries.' ... '[I now have] the enthusiasm to embark on teaching my India topic with greater background knowledge and confidence, having never taught about India before. [I also have the] awareness that teaching about India can be more cross-curricular than I originally thought.' ... 'I have gained many interesting literacy session ideas and I am completely rethinking geography.'

We have also received comments from teachers and children in schools who have recently used the pack. A selection is contained within the boxed section at the end of this appendix.

Summary

We are not alone in believing that *Thengapalli* is a unique resource, one that has already been used in hundreds of schools up and down the country. In the hands of creative teachers who can see its potential and share its vision, it can be a powerful inspiration and a force for a significant awakening. Through it, children can be made aware not only of their own responsibility for the environment, but also of their power to change things within it for the better. By seeing the problems that the people of Kesharpur encountered, getting into their skins and learning from their responses, they can appreciate the challenges that they faced, and which we all face. Furthermore, despite the disappointing fact that the pack has not contributed to local Agenda 21 actions, it is well placed to become a valuable resource for education for citizenship, which will be part of the National Curriculum from 2000.

One teacher's view

❦ We worked on *Thengapalli* in a block, every afternoon for an hour and a half, and it quickly took off, with ideas flowing. The class designed their own wooden *thenga*, which they ritually pass to one another, with the holder taking it home daily.

When we found out about eucalyptus trees, the class decided that they would like to plant a tree in the school grounds and wrote to the local Grounds Maintenance Officer, who came in and advised them on the best sort of tree to plant. They've also been able to offer advice to the local vicar on how to replace two trees that he lost.

As our school has a unit for children with moderate learning difficulties, who are integrated into many mainstream lessons, it was exciting to watch them join in and offer ideas which the other children listened to and acted on.

I found that it was an excellent way of raising the self-esteem of children with low self-esteem. It also gave children who have a narrow view of the world a much larger vision, raising issues of which they had no previous knowledge. It made them aware that they have the power to work together to change things. I am looking forward to teaching this again next year. ❧

(Veronica Lucas, Form 4 Teacher, Thornhill Junior School, Southampton)

The views of some 9-year-old children in her class

❦ The part I enjoyed most was when sides were arguing about whether the trees should be cut down. ❧
(Jareth Attard)

❦ Our class is going to get a tree to remember 4L (that's us). It's going to be a silver birch. It will be 2 metres when it comes and will grow to 10 metres. Trees are the lungs of the earth. ❧
(Emily Payne)

Appendix 1

❛ The most important thing we learned was ... how important trees are to us – they give us life and oxygen – and we learned not to cut down trees. ❜
(Jessica Wyse).

❛ I really liked the dancing when everybody was a different animal ... I liked it when I took the *thenga* home with me. Then I had to give it to someone else. You had to go around the circle and lay it in front of the person you want to give it to next. ❜
Daryl Bulmer).

We leave the last comment to an 11-year-old from Nightingale Junior School,[6] which tested many of the materials:

❛ *Thengapalli* isn't just an ordinary topic. It's not like geography, when you just go to the library and do a write-up. You find information for yourselves, not just in textbooks. Doing singing and dancing and drama you can enjoy it and learn about it, as you put yourself in other people's position. You can learn about culture and religion and lifestyles, and the difference. They cared about the future. It really happened, and they are still doing it. *Thengapalli* is about things that people don't always think about. It says "think of others" and "never give up on the environment" ❜

Appendix 2: BOJBP publications

1. BOJBP publications cover a wide range of topics:
   - Nursery raising and tree planting
   - Protecting the natural forest and hills
   - Water harvesting
   - Organic farming
   - Training farmers
   - Seed banks
   - Wildlife protection
   - Social forestry
   - Kitchen gardening
   - Herbal gardening
   - Networking
   - Sister organisations
   - Village unity and solidarity
   - Conflict management
   - Challenging traditions, e.g. dowry payments, early marriage
   - Atrocities against women
   - Uniting women for environmental protection
   - General health awareness
   - Birth control
   - Mother and child health
   - Environmental sanitation
   - Natural disasters

Appendix 2

2. The titles of some BOJBP books are given below.

- *Pracharapatra Panjika* (A compilation of leaflets published by BOJBP)
- *Pritipain Parichaya* (A description of 114 trees)
- *Asa Banjha Heba* (Environmental pollution and birth control)
- *Mo Sathi Katha* (Educational materials for children)
- *Chhunchi Chaluni Upakhyana* (Forest protection – action and reaction)
- *Chara Utariba Kipari* (How to raise a nursery)
- *Kasapita* (A collection of moral instructions)
- *Nisha Janta* (Fighting intoxicants)
- *Paribesh Sangeet* (A collection of songs written in the style of the epics)
- *Dubaganthi* (Detailed notes on herbal medicines)
- *Janile Hasihasaiba* (A description of useful plants)
- *Abandhu Jiban* (Friendships long and cherished)

Further information is available from BOJBP, PO Kesharpur (Buddhagram), Mandhatpur, Nayagarh, Orissa 752079, India.

Appendix 3: Update on BOJBP and the *Mahasangha*

The super-cyclone of October 1999

The good news is that villages in the 'mother area' of BOJBP were not affected by the super-cyclone that devastated large areas of coastal Orissa in October 1999, but many villages in areas where the *Mahasangha* has connections were affected. We discuss some of the *Mahasangha*-led rehabilitation work later in this Appendix.

Work in the mother area of BOJBP

In the 'mother area' of BOJBP no new activity has been initiated during the last year. Some inter-village conflicts arising from forest protection were resolved, but others were not. A conflict between Gamein and Gambhardihi over the felling of trees from the hill was amicably settled through the mediation of BOJBP. The conflicts between Badagarada and Sanagarada, and Gamein and Nagamundali were not resolved. Nevertheless, there has not been any adverse effect on Binjhagiri forest.

There is no significant change in the structure of the movement, except that the Executive Committee was increased from 25 to 34 members at the last General Body meeting on 9 January 2000. Three women now serve on the Executive Committee members.

Campaigning in new areas

During the last two to three years, campaigning work has shifted beyond the 'mother area' to the Odagaon block of Nayagarh District. Previously there had been several sporadic protest marches and meetings in Odagoan, but for the last 18 months campaigning efforts have been much more focused and continuous. Through a carefully devised strategy operating at the village and village-cluster levels, people have been made much more aware of the need for forest protection. As a result, the movement is now becoming strong in about 50 villages belonging to four *gram panchayats*: Odagaon, Ranganipatana, Nandighora, and Komanda, with which it had not had much earlier contact. The campaign has centred on three large stretches of

Reserved Forests covering three hills, namely, Gayalsingh, Bhalumundia, and Baghua. Sixteen villages of Odagaon and Nandighora *gram panchayats* have formed a Regional Committee, *Bhalumundia O' Gayalsingh Jangal Surakhya Parishad*. Similarly 14 villages of Ranganipatana *gram panchayat* now have their Regional Committee, *Gayalsingh Jangal Surakhya Parishad*. Another group of 16 villages have constituted *Baghua O' Gayalsingh Jangal Surakhya Parishad*. BOJBP is helping to build the capacity of these regional committees to protect the forest and resolve conflicts arising from tree felling and poor protection. A Central Committee has been formed by these Regional Committees, drawing up to five members from each region. This committee meets once every two to three months to review the protection of forest on the three hills. Because of this campaign, a sense of unity is reviving in many villages; for example, Ranganipatana, which has not held a whole village meeting for the last 100 years, is now able to hold fully inclusive village meetings.

BOJBP is also launching similar campaigns in the Kajalaipalli and Kesapania areas of Odagaon block, aiming to organise 28 villages in two groups to protect Ratnamala forest.

The women's movement

BOJBP has now organised 110 *Paribesia Mahila Surakhya Vahini* (PMSV: 'Women's Brigade for Environment Protection'), with a membership of 2100 women. These groups have collectively raised savings of Rs.900,000 for thrift and credit, and in the process have effectively freed themselves from the clutches of the local businessmen and money-lenders.

BOJBP has been trying to link these groups with the wider forest and environment protection movement. It has also been attempting to link them with financial institutions involved in micro-enterprise development, as there has been constant and consistent demand from the women's groups for financial support to scale up their economic activities. Negotiations are taking place with the local nationalised bank and with the National Bank for Agriculture and Rural Development. On 18 June 2000, a mass convention, attended by about 300 women, was organised to explore the possibilities for links with the banks.

Now women's groups in the Chaddhayapalli and Jagannath Prasad areas are actively involved in forest protection and income-earning activities, and also in collective decision making in their villages.

BOJBP and tribal development

BOJBP has been attempting to develop a plan of action for tribal development for a cluster of seven tribal villages in Barasahi *gram panchayat* of Odagaon block, Nayagarh District. So far it has organised village meetings and village-cluster meetings on forest and environment protection. One water-harvesting structure has been constructed, and two lift irrigation points have been installed for agricultural activities. This year the organisation is all set for a demonstration of organic farming in the area.

The *Mahasangha*

1999 was an important year in the history of the *Mahasangha*. During the first half of the year it evolved its new structure, based on four zones. Its network-building activity expanded to two more blocks of Nayagarh District, Dasapalla and Gania, which are very prone to conflicts and illegal logging by vested-interest groups. New members joined the *Mahasangha*. New leaders emerged, and efforts were made to create second-line leaders through participatory planning and exposure visits, both within and outside of Nayagarh.

The process initiated by *Mahasangha* in January 1997 for setting up a State-level forum on forestry culminated in the constitution of *Odisha Jangal Manch*, a State forum of forest-protecting communities, NGOs, and concerned individuals. The *Mahasangha* is the convenor of this forum, and as a part of its routine activities it became instrumental in promoting and facilitating networks of forest-protection committees in other districts. Now the members of the *Mahasangha* are attending or organising meetings in various parts of Orissa, including Bhadrak, Baleswar, Jagatsinghpur, Kendrapada, and Mayurbhanj Districts. Before the State Assembly elections this year, the State forum facilitated the preparation of a charter of communities' demands, centred mainly on forests and the environment. This charter was presented to all political parties contesting the elections, and also to the candidates. A State-wide campaign was launched to draw attention to the need for a comprehensive policy on non-timber forest produce, in order to make its management people-centred, with special emphasis on tribal and poor communities.

The *Mahasangha* is now initiating a campaign on specific issues relating to the rights of communities over forest resources. The emphasis is on needs-based extraction of fuelwood from forest areas that they protect, and the transit of fuelwood to neighbouring villages where people are short of fuel, in return for other products, in a kind of exchange of forest produce.

The *Mahasangha* is fighting to remove legal obstacles to this trade. Similarly for bamboo it is demanding that villages and individual bamboo-growers should have the right to use the produce and, if they want, to sell it. Then the paper and pulp industries could directly procure supplies from the communities or individuals.

After the cyclone, trees have fallen in some community-protected forests, and the Forest Department is minded to take that fallen timber from the villages. So there is now a campaign to sensitise people to challenge the action of the government. Efforts are also being made to influence the *gram panchayats* to place funds available for plantation and forestry development with the forest-protection committees. Once this is achieved, the forest-protection committees will have some kind of recognition from the government.

The *Mahasangha* is in a process of developing a curriculum for Middle Schools on environmental protection and management. Two rounds of workshops for teachers have already been organised for the implementation of this.

After the super-cyclone of October 1999, the *Mahasangha* and the State forum organised a campaign for the ecological rehabilitation of the cyclone-affected area. A team travelled through parts of Jagatsinghpur, Kendrapada, and Bhadrak Districts to make people aware of the need to protect trees and create forests. A team of experts in nursery raising from Nayagarh was sent especially to Kendrapada District, which had been particularly devastated, to train local people and organisations to raise nurseries and plantations. In this district, local organisations have received government support to raise one million seedlings. Seeds were also procured from the Nayagarh area and sent to the cyclone-affected areas. In Jagatsinghpur District, efforts have been made to constitute a 'district action group' of community-based organisations, small village-level groups, and NGOs to work towards the protection of environment. The focus of the campaign in this area is to raise plantations of multiple species, to create shelter-belt plantations, to revive mangrove forests, and to sensitise people concerning the community-based management of plantations and forests.

Manoj Pattanaik
July 2000

Notes

Chapter 1

1 'Scheduled' because they were listed in a schedule at the end of a law passed in 1936.
2 Although outlawed by the Indian constitution and subsequent legislation, 'bondage' is still widespread in India. People can be born into bondage (and live and die in it), or fall into it through debt to moneylenders, who may also be their employers. For very poor people it is easy to slip into bondage. Gita Mehta, quoting from the India School of Social Sciences in her book *Snakes and Ladders* (1997), writes: 'In a single northern village Farmer Minu had borrowed Rs.500 ($14) to buy food and was still working off the debt after 17 years. Farmer Nanda had borrowed Rs.100 ($3) for food and had already laboured without wages for twelve years. Farmer Sukha had worked for ten years for a loan of Rs.20 (75 cents) to buy a wooden box.'
3 'Loot' is a Hindi word.
4 Until recently Nayagarh was a sub-division of Puri District. In 1992/93 the Orissa Government redefined the old districts, of which there were 13, replacing them with 30 smaller districts, of which Nayagarh is one.
5 The common species are *Madhuca indica, Emblica officinalis, Acacia nilotica, Dalbergia paniculata, Aegle marmelos, Bahunia purpurea, Diospyros melanoxylon, Bridelia retusa,* and *Burea monosperma.*
6 Laterite blocks are rough-hewn from the thick encrustation which lies beneath the surface soil. This layer is not parent rock, but rather a concretised sub-soil.
7 All quotations attributed to Narayan and Subas Hazari are reprinted with permission from *Environmental Management in India*, Volume II, edited by P.K. Saparo, and published by Ashish Publishing House, New Delhi.
8 Bansidhar Sahoo, President of the village of Gamein and a member of the General Body of Friends of Trees and Living Beings, told us:
'We used to go to Sulia and Banijhari forests to collect firewood and timber for meeting our needs. These forests were located at a distance of about 10 km.

Those who had bullock carts used to go to Gochha forest, which is about 20 km – far from our village. People had to spend quite a lot of time for collection of firewood and timber.'

9   i.e. Binjhagiri. The people of Kesharpur sometimes call their village 'Buddhagram', meaning 'the village of enlightenment'. By the same token they call Binjhagiri 'Buddhagiri', meaning 'the hill of enlightenment'.
10 *Acacia arabica*.
11 There is some contention and rivalry concerning which village started forest-protection work first. Purna Chandra Pattanaik of Gambhardihi told us, '*Forest protection was not a new concept at that time, as Puania and Gambhardihi were [already] protecting their* gramya jangal [*village forest*].' This was confirmed by Baikuntha Pattanaik, a teacher in Kesharpur. On the other hand, Balabhadra Mohapatra of Manapur claimed that Binjhagiri village was '*the oldest forest protection village in the area*', and as such should have been given the international recognition that was later accorded the movement. Whatever is the truth of these competing claims, it likely that several villages had been doing this work for a number of years before they came together. It is also true that local village forest-protection work in Orissa goes back to the 1930s. (See Chapters 2 and 4.)
12 Jogi has adopted the title *shramik*, meaning 'worker'.
13 In fact we understand from recent conversations that in the end only Kesharpur ever followed this rule. All other villages strictly controlled where their goats grazed.
14 The Forest Corporation is the commercial arm of the Forest Department. It has depots in key villages, from which people can purchase fuelwood.
15 A *panchayat* is a group of villages clustered together to form the basic administrative unit. It is like a 'super parish council'.

Chapter 2

1   W. Fernandes, G. Menon and P. Viegas, 1988: *Forests, Environment and Tribal Economy: Deforestation, Impoverishment and Marginalisation in Orissa*, Indian Social Institute, New Delhi.
2   Government of India, 1998: *Enviro News*. Vol. 2, No. 1, Ministry of Environment and Forests, New Delhi.
3   D. N. Rao, 1992: 'Economics of forest products in India – issues and perspectives', in A. Agarwal (ed.): *The Price of Forests*, Centre for Science and Environment, New Delhi.

4 *Zamindaries* were the areas of land ruled over by *zamindars:* minor feudal landlords whom the British used for tax-collecting purposes.
5 Forest Survey of India, 1997: *State of Forests*, Ministry of Environment and Forests, Dehradun.
6 G. S. Padhi, 1985: *Forest Resources of Orissa*, Udyog Printers, Bhubaneswar.
7 SRUTI, 1995: *India's Artisans: A status report*, New Delhi.
8 J. B. Lal, 1992: 'Economic value of India's forest stock', in A. Agarwal (ed.): *The Price of Forests*, Centre for Science and Environment, New Delhi.
9 World Rainforest Movement, 1990: *Rainforest Destruction: Causes, Effects and False Solutions*, Penang, Malaysia.
10 Government of Orissa, 1996: *Agricultural Statistics of Orissa*, Directorate of Agriculture and Food Production, Bhubaneswar.
11 J.B. Lal, 1992: 'Deforestation: causes and control', in P. K. Khosla (ed.): *Status of Indian Forestry: Problems and Perspectives*, Indian Society of Tree Scientists, Solan.
12 Government of Orissa. 1994: *A Decade of Forestry in Orissa, 1981-90*. Chief Conservator of Forests, Orissa, Bhubaneswar.
13 Fernandes et al., op. cit.
14 Since the 1950s, successive Indian governments have undertaken central planning through the mechanism of five-year economic plans.
15 A. Mitra and S. Patnaik, 1997: 'Community Forest Management and the Role of Oxfam', Oxfam GB, Bhubaneswar.
16 One World Bank document, quoted in CSE 1985: *The State of India's Environment, 1984-85: The Second Citizens' Report*, Centre for Science and Environment, New Delhi, 1985, comments on the 'success' of farm forestry deriving from the 'spontaneous response of small farmers or communities to the commercial incentive of rising prices (for wood) and the perception that farming fast-growing short rotation trees as a cash crop can be as profitable as farming some of the traditional cash crops'. In case one should doubt the Bank's interest in the commercial aspect of social forestry, the report goes on to say: 'Marketing studies are seen as a critical issue in ... social forestry projects being funded by the Bank in India. It is becoming increasingly obvious that the rising prices for poles and fuelwood have been one of the main factors in stimulating farmers' interest in tree planting. Early definition of the size and geographic location of such cash markets can influence the siting of tree nurseries and the design of appropriate extension programmes.'

17 R. Chambers, N. C. Saxena and T. Shah, 1989: *To the Hands of the Poor: Water and Trees*, Intermediate Technology Publications, London.
18 M. Poffenberger et al., 1996: *Grassroots Forest Protection: Eastern Indian Experiences*, Asia Forest Network, Berkeley, USA.

Chapter 3

1 M. Poffenberger, B. McGean, and A. Khare, 'Communities sustaining forests in the twenty-first century', Chapter 1, in Poffenberger and McGean (eds.) 1996, *Village Voices, Forest Choices: Joint Forest Management in India*,, Oxford University Press, Delhi.
2 NCHSE, 1987: 'Documentation on Forests and Rights', Volume 1, National Centre for Human Settlements and Environment, New Delhi (mimeo), quoted in R. Chambers et al., 1989, op. cit.
3 It is reckoned that these 70 million comprise one-third of the world's total tribal/aboriginal population.
4 M. Gadgil and R. Guha, 1992: *This Fissured Land: An Ecological History of India*, Oxford University Press, Delhi.
5 W. Pereira and J. Seabrook, 1990: *Asking the Earth: Farms, Forestry and Survival in India*, Earthscan Publications, London.
6 M. Poffenberger, 'Valuing forests', Chapter 9, in Poffenberger and McGean (eds.) 1996, op. cit.
7 Verrier Elwin, cited in Gadgil and Guha, 1992, op. cit.
8 PRIA, 1984: *Deforestation in Himachal Pradesh*, Society for Participatory Research in India, New Delhi.
9 R. Chambers et al., 1989, op. cit.
10 Fernandes et al., 1988, op.cit.
11 In one factory in Orissa visited by Oxfam staff from Bhubaneswar, the number of unskilled labourers, who were mostly displaced tribal people, had declined from 10,000 to 1,000.
12 R. Chambers et al.,1989, op. cit.
13 In many States in India, forestry is the most important revenue-raising activity.
14 W.Fernandes et al., 1988, op. cit.
15 W. Fernandes and G. Menon, 1988, *Deforestation, Forest Dweller Economy and Women*, Indian Social Institute, New Delhi.
16 S. Ninan, 1981: *Women in Community Forestry*, included in the Report of the Seminar on the Role of Women in Community Forestry, Ministry of Agriculture and Co-operation, Government of India, Dehradun.
17 A. Mitra and S. Patnaik, 1997, op. cit.

18　An NSS survey quoted in K.P. Kannan and V. Vyasulu, 1983: 'An Alternative Strategy for Employment', Koraput Seminar, cited in Fernandes et al., 1988, op.cit.

Chapter 4

1　Quoted in R. Guha, 1989: *The Unquiet Woods: Ecological Change and Peasant Resistance in the Himalaya*, Oxford University Press, Delhi and University of California, Berkeley, cited in M. Gadgil and R. Guha, 1992, op. cit.
2　Compiled from (a) Guha 1989, op.cit.; (b) T. Rigzin, (ed.), 1997: *Fire in the Heart, Firewood on the Back*, Parvatiya Navjeevan Mandal, Uttar Pradesh; and (c) C. P. Bhatt, 1987: 'The Chipko Andolan – forest conservation based on people's power', in A. Agarwal (ed.): *The Fight for Survival*, Centre for Science and Environment, New Delhi.
3　*Uttarakhand* is the regional name of the northern part of Uttar Pradesh, comprising the hills and mountains of the Himalayan ranges.
4　*Sarvodaya*: a Gandhian movement dedicated to promoting community service and self-help.
5　*Appiko* is the Kannada word for 'hug'. (Kannada is the language of Karnataka.)
6　Compiled from Patrick McCully, 1996: *Silenced Rivers: The Ecology and Politics of Large Dams*, Orient Longman, New Delhi; S. Kothari and P. Parajuli, 1993: 'No nature without social justice: a plea for cultural and ecological pluralism in India', in W. Sachs et al. (eds.): *Global Ecology: A New Arena of Political Conflict*, Fernwood Publications, Canada.
7　B. Upadhyaya, 1994: 'People's Environmental Movements in Orissa', unpublished monograph.
8　This is the area that was so devastated by the 'super-cyclone' of 29–30 October, 1999.
9　Compiled from GASS, 1986. *Proposed Missile Base at Baliapal, Assault in the Name of Defence: A Report of the Fact-finding Team*, Bhubaneswar; Democratic Students Organisation, 1986: *Call of Baliapal*, Vijaywada, Andhra Pradesh; Unnayan 1990: *For Baliapal: A Handbook*, Calcutta.
10　Compiled from Update Collectives, 1993: *Save Chilika*, New Delhi; and Chilika Bachao Andolan and Krantidarsi Juba Sangam, 1993: *Maa Mati Chilika*, Orissa.
11　Compiled from A. B. Mishra, 1987: 'Mining a hill and undermining a society – the case of Gandhamarda', in A. Agarwal (ed.), 1987, op. cit.;

Records of Gandhamardan Surakshya Yuba Parishad; and personal discussion with Mr Nirmal, who was involved in the movement as a student activist.
12 P. Bal, 1998: 'Ama Odisa O Tata Company ra Sosana' (Oriya booklet), Gauli Vichar, Bhubaneswar.
13 S. Kant, 1990: 'Gandhian approach to the management of forest as common property resource: a case study of Binjhagiri Hill (Orissa) India', cited in Poffenberger and. McGean (eds). 1996, op. cit.

Chapter 5

1 N. Hazari and S. Hazari, 1989, op. cit.
2 Bahuguna's visit left a profound impression on many people in the movement. Suresh Chandra Swain of Gamein told us, *'When Sunderlal Bahuguna came to our village in 1983, we were really influenced by his action. We moved with him from village to village and at that time we were very much inspired. The area was also highly influenced [by his visit].'* Dr Hazari concurred. *'The protection of forests',* he told us, *'was strengthened by the visit of Bahuguna.'*
3 A block is an administrative sub-division of a district, defined for the purposes of government development programmes. In charge of a block is a Block Development Officer (BDO).
4 The movement has repeatedly challenged people on the issue of untouchability. But its impact was called into question by three of the women organisers in BOJBP, Mamata, Geeta, and Satyabhama, whom we interviewed for this book, who felt that *'the organisation has not addressed the question of untouchability properly'*. BOJBP has also campaigned against dowry (because of its drain on family resources), and against cremations (because of their demands on fuelwood supplies).
5 Throughout its life, the movement has concerned itself with traffic pollution and has occasionally campaigned on the issue. In August 1987 it had a postcard campaign to reduce 'sound pollution and road accidents'. Anyone who has travelled in India will know how much drivers of all vehicles, particularly of lorries, use their horns and will also know, in spite of that, how frequent accidents are, as testified by abandoned wrecks on the roadsides.
6 N. Baruah and S. Das, 1993: 'Issue Based Forestry Work and BOJBP', a paper presented at the Oxfam India Policy Forum, Lucknow.
7 While fining was not usually the preferred way of dealing with people caught breaking the rules, it was up to each village to decide on the most

appropriate punishment. Presumably in this case the people of Nagamundali felt that, since the culprit had actually stolen from the forest, a fine was appropriate restitution.

8 R. Davies, S. Chaudhuri, F. Rubin, P. Pradhan, A. Chaudhury, and S. Rath, 1994: 'An Evaluation of Two ODA Co-funded Oxfam Projects in Orissa, India', ODA, London. (ODA: Overseas Development Administration, part of what was then the UK Foreign Office. The Labour administration that came to power in May 1997 made ODA a separate ministry, known as the Department for International Development (DfID). Under the terms of the ODA/DfID annual block grant to Oxfam, a small sample of the work they co-fund is regularly evaluated. In 1993 BOJPB and one other Oxfam partner in Orissa, United Artists Association, were evaluated.)

9 S. Kant, N. M. Singh, and K. K. Singh, 1991: *Community Based Forest Management Systems (Case Studies from Orissa)*, Indian Institute of Forest Management, Bhopal, Swedish International Development Authority, New Delhi and ISO/Swedforest, New Delhi.

10 Ajay Mohanty, youth leader in Manapur, told us that in his village they had an aversion to fines. They preferred to require the guilty party to *'plant three banyan trees and nurture them until they became big'*.

11 *Satyagraha*: non-violent truth force or protest. *Dharna*: a sit-in.

12 Having received notification of the award, they were invited to Brussels to receive it. But regarding the cost of two airfares to Europe as excessive, they decided to spend the money on equipping 4000 households with pit latrines. This was part of their environmental health programme.

13 N. Hazari and S. Hazari. op. cit.

14 Mitra and Patnaik, 1997, op. cit.

15 Few people in these villages are wealthy. 'Better-off' is a more accurate term than 'rich'.

Chapter 6

1 Mitra and Patnaik, 1997, op. cit.

2 Recently it has been decided that, since the number of unresolved conflicts is on the increase, the sister organisation in whose area the conflict occurs will first try to reconcile the parties. If the sister organisation is unsuccessful, the dispute will then go to a zonal-level body. These have yet to be formed. If the conflict cannot be resolved at that level, the specially constituted sub-committee of the *Mahasangha* will take it up.

3 Village commons are the common-property resources of a village, owned collectively by the community. They include patches of forest, grazing land, community ponds (tanks), graveyards, temples, and village halls.
4 See Chapter 5 for Dr Hazari's position on resolving internal disputes internally and not running to outsiders for help. This stricture has been ignored in recent years, when on more than one occasion the Forest Department has been asked to intervene in disputes between different villages.
5 That is all the villages of the sister organisations, which were the sister organisations of BOJBP, plus others which have joined recently, together with single villages which have their own committees – independent village forest-protection committees (IVFPCs).

Chapter 7

1 S. Kant et al., 1991, op. cit.
2 'Sanskritisation' is the process whereby non-Hindu groups and communities (such as tribals) adopt the norms and values of Hindu India, including caste hierarchy.
3 No other NGO or CBO in India has such a long and continuous funding relationship with Oxfam.
4 A large proportion of this £30,000 has been provided on a 50:50 co-funding basis with the UK government's Department for International Development.

Chapter 8

1 N. Hazari and S. Hazari, 'Environmental revolution in Orissa: the Buddhagam experiment', in R. Atkar, 1990: *Environmental Pollution and Health Problems*, Ashish Publishing House, New Delhi.

Appendix 1

1 The UK government's National Curriculum divides children's education into four age-related stages, known as Key Stages. Key Stage 2 is from 8 to 11. Within each age group, levels of attainment in knowledge, skills, and understanding are tested by Standard Assessment Tests (SATs). If resource materials are to have any chance of commercial success, they must fit the curriculum needs of the stage at which they are aimed, otherwise teachers are not likely to use them.

2 'Agenda 21', which emerged from the Rio Earth Summit in 1992, is an environmental action plan for the twenty-first century, agreed by 175 countries, including the UK. It assumes that governments alone cannot solve problems, and it challenges local authorities to produce an Agenda 21 Action Plan for their areas, in partnership with local people, businesses, and other organisations. The 'futures dimension' is about creating a vision and plan for a sustainable future: thinking globally, acting locally.

3 K. Grimwade, 'India in action', *Times Educational Supplement*, 27 March 1998.

4 The workshop was organised by the Canterbury-based World Education Development Group in November 1998.

5 The cost is £26.50 per pack, plus £6.30 for postage and packing. The video costs £20.00. Both are obtainable from Hampshire County Council (HIAS, Education Department, Country Office, The Castle, Winchester, SO23 8UG, UK). The pack consists of a teacher's overview booklet, a locality-study booklet, a full-colour A3 village map, a booklet about sustaining environments, a colour photopack (with 105 photographs), and an audio cassette (with songs, sound trails, and music from Orissa for dancing).

6 Quoted in an article by Helen Roberts, 1998: 'Music of time and place: songs of forest and community', in the summer issue of *Music Today*. Helen Roberts was one of the Hampshire LEA advisers who went on the Kesharpur study visit in 1995; she wrote the music component of the pack.

Further reading

AFC, 1994, *People's Participation in Forestry – Joint Forest Management*, Agriculture Finance Corporation Ltd., New Delhi.
Anderson, R. and W. Huber, 1988, *The Hour of the Fox: Tropical Forests, The World Bank and Indigenous People in Central India*, Vistaar Publications, New Delhi.
Arnold, J. E. M., 1991, 'Learning from farm forestry in India', *Forests, Trees, and People* Newsletter 13: 43-6.
Arora, D., 1994, 'From State regulation to people's participation: case of forest management in India', *Economic and Political Weekly* 29(12): 691-8.
Chambers, R., N. C. Saxena, and T. Shah, 1989, *To the Hands of the Poor: Water and Trees*, Intermediate Technology, London.
Chatterjee, M. (undated), *Women in Joint Forest Management: A Case Study from West Bengal*, Technical Paper 4, IBRAD, Calcutta.
Chopra, K., G. K. Kadekodi, and M. N. Murthy, 1989, *Participatory Development: People and Common Property Resources*, Sage Publications, New Delhi.
Conroy, C. et al., 1999, *Learning from Self-Initiated Community Forest Management Groups in Orissa: Practices, Prospects and Policy Implications*, a paper on the study conducted by NRI, UK.
Dhar, S. K., J. R. Gupta, and S. Madhu, 1991, *Participatory Forest Management in the Shivalik Hills*, Sustainable Forest Management Working Paper No.5, Ford Foundation, New Delhi.
Gadgil, M. and R. Guha, 1992, *This Fissured Land: An Ecological History of India*, Oxford University Press, Delhi.
Guha, R., 1991, *The Unquiet Woods: Ecological and Peasant Resistance in the Himalayas*, Oxford University Press, New Delhi.
Hazari, N. and S. Hazari, 1989, 'Community action in environmental conservation – an experiment in Orissa', in P.K. Saparo (ed.), 1989, *Environmental Management in India, Volume II*, Ashish Publishing House, New Delhi.
Hazari, N. and S. Hazari, 1990, 'Environmental revolution in Orissa: the Buddhagram experiment', in R. Aktar, 1990, *Environmental Pollution and Health Problem*, Ashish Publishing House, New Delhi.

Hazari, N. and S. Hazari, 1991, 'Lessons of the Buddhagram environment movement', in I. Mohan, 1991, *The Fragile Environment*, Ashish Publishing House, New Delhi.

Jonsson, S. and A. Rai, 1994, *Forests, People and Protection: Case Studies of Voluntary Forest Protection by Communities in Orissa*, ISO/Swedforest, New Delhi.

Kant, S., N. M. Singh, and K. K. Singh, 1991, *Community Based Forest Management Systems (Case Studies from Orissa)*, Indian Institute of Forest Management, Bhopal, Swedish International Development Authority, New Delhi, and ISO/Swedforest, New Delhi.

Mahapatra R., 1999, 'On the warpath', article in *Down to Earth*, 30 September, Centre for Science and Environment, New Delhi.

Mitra, A. and S. Patnaik, 1997, 'Community Forest Management in Orissa and the Role of Oxfam', Oxfam GB, Bhubaneswar.

Operations Research Group, 1985, 'Case Studies on Rehabilitation of Degraded Forests', sponsored by Directorate of Social Forestry Project, Orissa, Bhubaneswar (mimeo).

Index

adivasi (original people) 5, 38–40
afforestation 12, 26, 47
   compensatory 26, 28–9, 31–2
   see also reforestation
Agenda 21 150, 167–8n2
agriculture 3, 66
   farm forestry 32–3
   forest dwellers 24–5
   organic farming 14, 91
   shifting cultivation 26, 29–30, 38, 40–6, 52
Anchalik Sulia Bana Surakhya Committee 116–17
animal husbandry 42
Anti-Hirakud Dam 64–5
Anti-missile Base Committee 66–7
Appiko 61–2, 164n5
aquaculture 67–8
artisans 24, 45
avenue plantation 31
awareness-raising 75–6, 118, 132, 133
Awasthapada 17, 136

babul trees 13–14
Badagarada 109–10, 111–12, 156
Bahuguna, Sunderlal 60–1, 70, 72, 74, 137, 165n2
Baliapal National Test Range 66–7
bamboo 49, 53, 104, 112, 159
Banamahotsava (Forest Festival) 76–7, 91, 100
Basundhara Bandhu 106
Beguniapatana 11
benefit-sharing 111, 125, 130, 138, 140
Bharat Aluminium Company (BALCO) 68–70
bilateral donors 140–2, 144
Binjhagiri
   deforestation 3, 11, 12
   forests 7
   regenerated forests 135
   wildlife 3
   women's forest work 109
biotic pressures 30, 36
Britain
   DfID 140, 167n4
   National Curriculum 150, 167n1
   Thengapalli Development Education Project 8, 136, 145–53, 168n5
Bruksha O' Jeevar Bandhu Parishad (Friends of Trees and Living Beings) 1, 16–17, 125–6

agencies responsible 17–21
awards 93, 137
benefit-sharing 111
caste 129
challenges 73–4
community conflict 97, 109–10, 156
Community Forest Management 35, 37, 137, 139
cultural projects 77–9, 99
dowry system 79–80
environmental/human issues 93, 96
governance 88–90
Joint Forest Management 115–18, 137, 138
leadership 17–18, 21, 93, 124–5, 127–8, 142, 143–4
and *Mahasangha* 101–2, 122–4, 126
Oxfam 75, 81–2, 84–5, 90–1, 95–6, 119, 136, 142, 144
Oxfam–ODA evaluation 86, 98, 106, 142
publications 78–9, 154–5
regional committees 157
school children 10, 19, 80, 132–3
sister organisations 80–1
Social Forestry 113–15
staff 118–20, 125
tribal development 100, 158
trust in 74, 111–12, 120
villages 6, 88–9
voluntarism 118–20
Women's Development Programme 83, 105–7, 110–11, 124–5
women's participation 87–8, 99, 110–11
World Environment Day 76

cash-for-work scheme 104
caste system
   BOJBP area 129
   Forest Caste 100–1, 129, 135
   Jatiana Sabha 100–1
   Scheduled Castes 3, 5, 10, 40
Chambers, R. 33, 50, 53, 54
charcoal 45
children: see school children
Chilika Bachao Andolan (Save Chilika Movement) 68
Chipko movement 59–62, 74

Community forest management

civil resistance 74
class 95–6, 129
   see also caste
clear-felling 26, 36, 47, 49
coastal environment 65–8
commercialisation 26, 48, 55
communal action 129–30
community conflicts
   BOJBP 97, 109–10, 156
   deforestation 111–12
   Mahasangha 97, 166n2
   resolution 99, 103, 141
Community Forest Management 36, 137, 139
   BOJBP 35, 37, 137, 139
   Joint Forest Management 34–5, 36
   Mahasangha 137, 139
   Oxfam report 33
community woodlots 33
commutation fees, forest produce 112
compensation, lack of 52, 64–5
compensatory afforestation 26, 28–9, 31–2
credit schemes 99, 106–7, 121, 132, 157
cultural programmes 77–9, 99

dalits (oppressed) 3, 5, 76, 95, 96, 132
dams
   Narmada River 52, 62–3
   non-government organisations (NGOs) 65
   Orissa 52, 63–5
   women 65
Dashauli Gram Swaraj Sangh (DGSS) 59
debt 40, 48, 54, 55, 160n2
deforestation 45
   Binjhagiri 3, 11, 12
   BOJBP 93
   community conflict 111–12
   exploitation 22, 28, 46–51
   fuelwood 54–5
   herbal medicines 55
   and regeneration 7
   shifting cultivation 29–30
   tribal peoples 5
   women 54–5
development
   class 95–6
   coastal environment 65–8
   projects 51–2
   tribal people 121
DfID (Department for International Development) 140, 167n4
dharna (sit-ins) 83–4, 93, 166n11
displacement
   compensation 52
   forest dwellers 52
   mechanisation 53
   National Test Range 66–7
   Orissa 52–3, 63–5

shifting cultivation 52
   tribal peoples 40
dowry system 79–80, 132, 165n4
drought-relief programme 104, 130

economic liberalisation 53
encroachment 28–9, 36, 45–6, 48, 55
Environment and Forests Central Ministry 35–6
environmental degradation 15, 47, 67–8
environmental education 77–8, 91, 99, 133
environmental movement
   activism 144
   gender equity 105–6
   pro-poor 58
   sacred quality 74, 76
environmental protection 14, 16, 58–9, 107, 111–13
exotic trees 50, 114, 140
exploitation 5, 54, 57
   deforestation 22, 28, 46–51

fallow system 41, 48
farm forestry 32–3
fasts as protest 60–1, 83–4
felling
   illegal 30–1, 56, 108, 111
   moratorium 61
   see also clear felling
fines 84, 165–6n7
foods 24, 42–3, 55–6
Forest Act 25–6
Forest Caste 100–1, 129, 135
forest conservation: see forest protection
Forest (Conservation) Act 26, 27–8, 31
Forest Department
   cash-for-work scheme 104
   Joint Forest Management 34–6, 117–18
   Orissa 23–4, 28
   Range Officers 20–1
   woodlots 33
forest dwellers
   displacement 52
   farmers 24–5
   gatherers 38
   gender 54–5
   knowledge 41–2
   non-timber forest produce 49–50
   policy makers 37, 139–40
   progress 46–51
   rights 27, 28, 51, 53–4
   shifting cultivation 38, 40–6
   spirituality 45
forest management 7, 47
   community-based 116–17
   experts/local people 140–1
   Orissa 27–8, 71
   see also Community Forest Management; Joint Forest Management; Social Forestry

172

Index

forest produce 50
  commutation fees 112
  foods 24, 42–3
  grazing 26, 36
  marketing 100, 114–15, 141
  Orissa 71
  subsistence 47–8
  see also non-timber forest produce
Forest Produce Marketing Initiative 114–15
forest protection 161n11
  campaigns 99
  commercial interests 26
  regulation 113
  school children 19
  women 108
forest regeneration 7, 14, 31–2, 133, 135, 140
  see also afforestation; reforestation
Forest Survey of India (FSI) 23, 30
forests
  biotic pressures 30, 36
  communities 25–7
  designations 7, 22–4, 26
  eco-systems 22
  economic/social functions 24–5
  exotic/native trees 14, 50, 114, 140
  leasing 26, 60
  marketing 100, 114–15, 141
  revenues 163n13
  usufruct rights 26, 28, 103–4
  see also deforestation; Protected Forests; Reserved Forests
Free Narmada Movement 63
Friends of Trees and Living Beings: *see Bruksha O' Jeevar Bandhu Parishad*
fuelwood
  babul trees 14
  deforestation 54–5
  quality 43
  shortage 15, 30, 130
  subsistence gathering 11, 13, 36, 158–9, 160–1n8
  urban dwellers 25

Gambhardihi 18, 84, 156
Gamein 11, 83, 156
Gana Sangram Samiti (Mass Movement Committee) 70–1
Ganatantrik Adhikar Surakhya Sangathan (GASS) 67
Gandhamardan aluminium industry 68–70
Gandhian tradition
  dharna 93
  forest regeneration 14
  non-violence 72, 95, 128, 131, 143
  resistance 74
  satyagraha 93

gender
  decision-making 55
  environmental movement 105–6
  equity 105–6, 110–11, 143
  forest dwellers 54–5
  health 3, 55
  integration 98, 110–11
  Joint Forest Management 35
  tribal people 54–5
  see also women
gender training 87–8, 143
goat-keeping 12, 13, 130, 161n13
Gopalpur steel plant 70–1
government agencies 20–1
grazing 26, 36
  see also goat-keeping
green culture 134–5, 144
Green Revolution 42

habitat destruction 5, 7, 22
Hampshire LEA 146, 148
Hazari, Dr Narayan
  Balia incident 11–12
  BOJBP 16
  community conflicts 94–5
  community projects 17
  fasts 83–4
  Forest Department 20–1
  goat-keeping 13
  Kesharpur factionalism 10
  links with external agents 127, 135–6
  local participation 139
  Malati Hill saplings 14
  padayatra 76
  social activism 18–19
  stone quarrying 73–4
Hazari, Subas 11–12, 17, 136, 139
health 3, 55–6, 87, 100, 107
herbal medicines 42, 49, 55, 56
Hinduism 80, 167n2
Human Environment Conference, Stockholm 58
human resources 98–9
hunger strikes 108
hunter-gatherers 40
hunting 38, 73, 82–3

Indian Constitution 40
Indian Forests Charter 25
Indian Social Forestry Programme 32–4
Indravati Gana Sangharsha Parishad 65
industrial projects 26
infrastructure projects 26, 51–2
irrigation schemes 52
IVFPCs 81, 92–3, 94, 99, 167n5

Jalsamadhi (mass suicide by drowning) 65
Jatiana Sabha (caste group) 100–1

173

Jena, Susanta 18, 120, 127
Joint Forest Management
  BOJBP 115–18, 137, 138
  Community Forest Management 37, 137
  Environment and Forests Central Ministry 35–6
  Forest Department 34–6, 117–18
  local communities 34–6
  Mahasangha 137, 138
  Social Forestry 104
  women's participation 35

Kesharpur 7–11
  pilgimage 74
  schools 10, 129
  unity 129–30, 131
Khatei, Udayanath 12, 16, 17–18, 21
Khesra (Undemarcated Protected Forests) 7, 103–4

labour
  bonded 40, 160n2
  seasonal migrant 10, 55
land rights 52
leadership
  BOJBP 17–18, 21, 93, 124–5, 127–8, 142, 143–4
  protests 62–3
  Sahoo 15
leprosy 77, 132
livelihoods 98
local communities: see community projects
logging 25, 61–2, 158
  see also felling

Mahamandal 106
Mahasangha 94, 102–5
  BOJBP 101–2, 122–4, 126
  campaigns 99, 100
  community conflicts 97, 166n2
  Joint Forest Management 137, 138
  Oxfam 136, 142
  Sahoo 120–1, 123
  Social Forestry 113–15
  super-cyclone 156
  usufruct rights 140, 158–9
  women's networks 103, 105
  women's participation 109, 111
mahila mandal (women's group) 60
mahua plant 43, 49
Malati Hill saplings 14
Malatigiri forests 7, 11, 135
Mandal 106
marketing of forests 100, 114–15, 141
mechanisation/displacement 53
migrant labour 10, 55
mines 52
Mishra, Professor A. B. 69

Nagamundali 156
Narmada Bachao Andolan (Free Narmada Movement) 63
Narmada River 52, 62–3
National Centre for Human Settlements and Environment 38, 39
National Commission on Agriculture 26
National Forest Policy 26, 27, 34–6
National Service Scheme 12, 13, 16, 20, 74, 135–6
National Test Range, Baliapal 66–7
nationalisation
  non-timber forest produce 28, 50–1
natural resources
  exploitation 5, 54, 57
  human survival 58
  needs-based extraction 158–9
  poverty 96–7
Nayagarh 10
Nayagarh Jangala Surakhya Mahasangha 82, 94, 97, 98, 100–5
  see also Mahasangha
non-government organisations (NGOs) 58, 65, 143–4, 144
non-timber forest produce
  commercialisation 49–50, 55
  forest dwellers 49–50
  nationalisation 28, 50–1
  poverty 158
  pressures on 48
  subsistence gathering 24–5
  usufruct rights 34
non-violence 72, 95, 128, 131, 143
  see also satyagraha
Nuagaon festival 78
nurseries 80, 85–6, 91, 99–100

ODA Report 87, 95
Odisha Jangal Manch 105, 137, 158
organic farming 14, 91
Orissa 3–6
  avenue plantation 31
  coastal environment 65–8
  dams 52, 63–5
  Department of the Environment 85
  displacement 52–3, 63–5
  Forest Department 23–4, 28
  Forest Development Corporation 104, 161n14
  forest management 27–8, 71
  forest produce 71
  Ganatantrik Adhikar Surakhya Sangathan 67
  Gopalpur steel plant 70–1
  Oxfam project report 85
  Protected Forests 27–8, 29
  Save Gandhamardan Movement 68–70
Orissa Forest Act 27
Oxfam
  Asia Desk 149

Index

BOJBP 75, 81–2, 84–5, 90–1, 95–6, 119, 136, 142, 144
   capacity building 143
   Community Forest Management report 33
   credit schemes 106–7
   criticised 93, 121–2, 142–3
   Education Department 146, 148
   gender training 87–8, 143
   Mahasangha 136, 142
   Project Officer 87, 92–3
   Project Report on Orissa 85
   reports 55
Oxfam Bhubaneswar 5, 149
Oxfam India 47
Oxfam–ODA evaluation 86, 98, 106, 142

padayatra (campaigning march) 75–6, 131–2, 135
panchayats (groups of villages) 50, 51, 161n15
paper industry 25, 32, 49, 53, 60, 113, 139
Paribesia Mahila Surakhya Vahini (Women's Brigade for Environment Protection) 106, 107, 157
pastoralist workshops 100
Patnaik, Pratap 12, 13, 20, 137
persuasion 82, 84, 131, 143
plantations 31, 47, 56, 100, 133, 140
pokasunga shrub 13
policy makers 139–40
Population Foundation of India 3
postcard campaigns 91
poverty
   BOJBP 93
   environmentalism 58
   exploitation 5
   livelihoods 98
   natural resources 96–7
   non-timber forest produce 158
   Orissa 3
   protection 95, 96
   representation 131
   Social Forestry Programme 51
   tribal people 38–40
   usufruct rights 130–1
Prakash, Bhagaban 16
prawn culture 67–8
prostration 82–3, 94, 108, 110
Protected Forests 23, 24, 26, 53, 103–4
   Orissa 27–8, 29
protests 62–3, 72, 82–3, 110
Puania 84
publications, BOJBP 78–9, 154–5
puja (prayerful devotion) 80, 81
pulp industry 32, 139
punishment 131, 165–6n7, 166n10

Radhamohan, Professor 16, 18–19, 85, 86
Ramsar Convention 67–8

Range Officers 20–1
Ratnamala Jungal Surakhya Committee 106, 109, 112
reforestation 7, 11–12, 14
   see also afforestation
Rehabilitation of Degraded Forests 31
Reni forests 59–60
Reserved Forests 23, 24, 26, 27, 53, 103–4
resin 44, 60
rice growing 42
rights for forest dwellers 27, 28, 51, 53–4

Sabuja Jeevan 106, 109, 111
Sabujima, Mahasangha newsletter 117
Sahoo, Bhagaban 78
Sahoo, Joginath 12, 161n12
   BOJBP 21, 73
   fasting 84
   leadership 15, 129
   Mahasangha 120–1, 123
   Malati Hill saplings 14
   prostration 82–3, 94
   resignation 120–1
   school children 133
   social activism 18–19, 127
   song 78
   tree planting 74
   visit to Britain 149
   wedding ceremony 79
   wildlife protection 73
sal seeds 49, 50
Sanagarada 111–12, 156
sanitation 87, 100, 107, 132
Sanskritisation 167n2
saplings 79, 85–6, 134
Sarvodaya leaders 61, 164n4
satyagraha (non-violent protest) 93, 110, 166n11
Save Chilika Movement 67–8
Save Gandhamardan Movement 68–70
Scheduled Castes 3, 5, 10, 40
Scheduled Tribes 3, 5, 38–40
school children
   BOJBP 10, 19, 80, 132–3
   Britain 136
   environmental activism 144
   nurseries 80
   plantations 133
schools, Kesharpur 10, 129
seasonal migrant labour 10, 55
seed bank 78–9, 85, 86–7, 100, 140
seed sowing 80–1
seedlings 32, 33, 115
settled cultivators 45–6
shifting cultivation 29–30
   displacement 52
   forest dwellers 38, 40–6
   National Forest Policy 26

175

Community forest management

SIDA (Swedish International Development Authority) 140
sit-ins *(dharna)* 83–4, 93, 166n11
smuggling of timber 30–1
  see also felling, illegal
social activists 18–19
Social Forestry 32–4, 51, 100
  BOJBP 113–15
  Joint Forest Management 104
  Mahasangha 113–15
  World Bank 162n16
social justice 40, 72
soil erosion 14, 48, 94
songs 77–8, 134–5
spirituality 45, 80, 134–5
steel plant 70–1
Stockholm Conference on Human Environment 58
stone quarrying 73–4, 84
subsistence 24–5, 47–8, 111
  fuelwood gathering 11, 13, 36, 158–9, 160–1n8
  hunting 38, 73
suicide, mass 65
Sulia Bana Surakhya Committee 112
Sulia Paribesh Parishad Anchalik 112, 140
sustainability 40–6, 48, 111–13
Symonds & Co. 59

Talapatana women 107–8
Tatas 67–8, 70–1
teachers 18–19, 99, 159
tendu trees 49, 50
Thengapalli Development Education Project 8, 136, 145–53, 168n5
thengapalli (forest protection rota) 13, 89, 108
timber 30–1, 51–2, 113–15
  see also fuelwood
timber harvest 14, 25, 30, 34
tree planting 20, 74, 79, 134–5
tribal people
  BOJBP 100, 158
  deforestation 5
  development programme 121
  displacement 40
  gender roles 54–5
  knowledge 41–2
  poverty 38–40
  settled cultivators 45–6

Unclassed Forests 23
Undemarcated Protected Forests 7
United Nations Environment Programme 93, 137
untouchability 165n4
  see also dalits
Upper Indravati dam 65
usufruct rights 111
  benefit-sharing 140

forests 26, 28, 103–4
  Mahasangha 140, 158–9
  non-timber forest produce 34
  poverty 130–1
  Protected Forests 28
  Reserved Forests 103–4
Utkal University 16, 20, 68, 135–6
Uttarakhand Sangharsh Vahini (USV) 60–1, 72

Vana Sanrakhyana Samitis (Forest Conservation Societies) 34
village commons 166–7n3
Village Councils 88, 89, 131
Village Forest Committees 51
Village Forest Protection Committees (VFPCs) 34, 125
  independent 81, 92–3, 94, 99, 167n5
Village Forests 23, 26
violence against women 107
voluntarism 91, 118–20, 128, 129–30
voluntary agencies 20
volunteers 98–9, 125
wasteland development 33–4
wildlife
  Binjhagiri 3
  environmental degradation 15
  hunting 38, 73, 82–3
  protection 93, 99
  reforestation 14
Wildlife Conservation Week 77, 91
wildlife sanctuaries 52
women 88
  credit schemes 99, 106–7, 121, 132, 157
  dam protests 65
  deforestation 54–5
  forest protection 108
  forest work 109
  protests 62–3, 83, 110
  violence against 107
  see also mahila mandal
women's associations 88, 103, 105
Women's Development Programme
  BOJBP 83, 105–7, 111, 124–5
women's participation
  BOJBP 87–8, 99, 110–11
  Joint Forest Management 35
  Mahasangha 103, 105, 109, 111
woodlots 33, 139
World Bank 62, 65, 140, 162n16
World Environment Day 76, 91
World Forest Day 91
World Health Day 77

youth volunteers 99, 125

zamindaries (land divisions) 23, 24, 162n4